Volume 7
Cybernetics and Systems Series

DECISION MAKING ABOUT DECISION MAKING: Metamodels and Metasystems

DECISION MAKING ABOUT DECISION MAKING: Metamodels and Metasystems

edited by

John P. van Gigch

California State University

With a Foreword (Metacomment) by

Stafford Beer

Contributors

John P. van Gigch, Walter J. M. Kickert, C. O. François,
Magoroh Maruyama, Burton H. Voorhees, Maria Nowakowska,
Vladimir A. Lefebvre, John R. Battista, Ian I. Mitroff,
Norman A. Coulter, Jr., Albert L. Johnson, Rammohan K. Ragade,
Mildred L. G. Shaw, Gerard de Zeeuw, Ranulph Glanville,
Werner Ulrich, Ann F. van Gigch, Nora Femenia.

ABACUS PRESS

Cambridge, Mass

&

Tunbridge Wells, Kent

First published in 1987

ABACUS PRESS
PBS, PO Box 643,
Cambridge,
Mass 02139, USA

and

ABACUS PRESS
Abacus House,
Tunbridge Wells,
Kent, TN4 0HU, UK

The Cover

Pascal Lefebvre is a young French artist who claims to use surréalisme as a basis to construct his canvasses. This painting is titled 'Metamodels and Metasystems' and is therefore particularly appropriate.

British Library Cataloguing in Publication Data

Van Gigch, John P.
Decision making about decision making :
metamodels and metasystems. — (Cybernetics
and systems series v.7)
1. Decision-making 2. System theory
I. Title II. Series
658.4'032 HD30.23

ISBN 0-85626-200-5

Printed in Great Britain by
Antony Rowe Ltd., Chippenham

CONTENTS
by classification and relevance

	Epistemology & Philosophy of Science	Systems Theory & Methodology	Research Methodology & Inquiring Systems	Psychology	Cybernetics & Control Theory	Management Science & Operations Research	Organization Theory & Decision Making	Knowledge & Expert Systems	Social System Design	Organizational Dysfunctions & Systems Failures	Health-Care Administration	Information & Communication Theory	Logic & Hierarchy Theory	Modelling & Metamodelling
INTRODUCTION														
Overview van Gigch	●	●	●	●	●	●	●	●	●	●	●	●	●	●
Ch. 1 van Gigch	●	●								●				●
Ch. 2 van Gigch	●	●	●			●	●							●
THE METASYSTEM PARADIGM														
Ch. 3 Kickert & van Gigch	●	●	●		●	●				●				
Ch. 4 van Gigch	●	●												●
Ch. 5 François	●	●			●									●
EPISTEMOLOGIES														
Ch. 6 Maruyama	●						●		●			●		●
Ch. 7 Voorhees	●	●		●										
Ch. 8 Nowakowska	●							●	●				●	●
Ch. 9 Lefebvre		●		●										●
Ch. 10 Battista	●		●	●								●		●
Ch. 11 Mitroff & Mason		●	●				●					●		
Ch. 12 Coulter & Johnson		●			●									●
METAMODELS & APPLICATIONS														
Ch. 13 Ragade		●	●			●		●				●		●
Ch. 14 Shaw		●		●				●						●
Ch. 15 de Zeeuw			●						●				●	●
Ch. 16 Glanville							●		●					●
Ch. 17 Ulrich							●		●				●	
Ch. 18 van Gigch		●			●					●		●	●	●
Ch. 19 van Gigch		●			●					●		●	●	●
Ch. 20 A. F. van Gigch		●							●		●			●
Ch. 21 Femenia		●		●						●	●			●
CONCLUDING REMARKS														
Ragade	●	●	●											●

CONTENTS

FOREWORD

by Professor Stafford Beer

METACOMMENT

How does the brain work? It would be nice to know. Aristotle noted all the convolutions in the cortical surface, and said that this was a system for cooling the blood. Descartes, aware of contemporary hydraulic technology, thought of the brain in terms of 'the fountains in the King's gardens'. Technology had changed by the time of Locke to the strongly mechanical: he postulated invisibly thin wires running in the tubes of nerve processes. Then came electrics; and the brain was a kind of telephone exchange. Today, of course, because we are still victims of our own current technology, the brain 'is a hologram'.

Where in particular has forty years of neurocybernetics led? The doyen of the field, the late Warren S. McCulloch, described the brain as a three pound electro-chemical computer, running on glucose at twenty-five Watts. He went on to develop models[1] of the workings of the neuron, and of assemblies of neurons. Now the individual neuron receives such a vast number of excitatory and inhibitory inputs that it is tempting to think of it as an analogue device. McCulloch said that the transfer function is a seventh order differential equation. (Not everybody knew that this was because he could get only seven probes into a Purkinje cell.) But certainly the neuron has a *digital* output. Either the axon that emerges from the neuron fires, or it doesn't. Thus brain models tend to operate in binary logic, and their descriptions to employ Boolean algebra.

From this basis, I myself came to entertain the hypothesis that a brain so constituted might have no option but to *dichotomize* any issue it addressed. We would end up with good and evil, black and white, and so on. Then degrees of indifference, shades of grey, would be intellectual constructs trapped along a scale closed by polar opposites at each end. It looks like that in practice, as we hear fundamentalists rave in religion, in morality, in politics. For many years, then, I tried to prove this neurocybernetic postulate about the brain. The nearest I came to a demonstration preserved in print was in a critique of a research machine called the Perceptron made at a conference in 1958[2]. In subsequent personal discussions, McCulloch airily said, "yes, of course"; but it meant that the brain must use at least a *ternary* logic. And this from the architect of binary neural nets!

This comes down to speculating about a physical basis for the problems generated by our humanity. For if brain operations are well described in binary logic, then we might have an *in-built* propensity to dichotomise, and the brain will need at least a third term over and beyond its system of polarities to make sense of the world. The Greek word for 'over and beyond' is *meta*.

When the story is followed into theoretical logic, it loses its speculative character and becomes a serious business. Polarizing concepts, the A and the non-A, result in paradoxes of many kinds. 'This statement is false'. If it is, it isn't, and if it isn't, it is'. Even so, paradoxes can be resolved. But they lead on into the region of absolute undecideability, where Gödel's Theorem about the incompleteness of formal logics undermines the nature of reasoning itself. The escape from these problems always lies in logical *recursion*: languages have to be devised that are 'over and beyond' the language in which we happen to be trapped and these are therefore called metalanguages. Metamodels and metasystems come along by extension, because models and systems are both logical structures that exhibit special properties in the mode of their mapping what is empirically presented, and fall foul of exactly similar logical traps.

Now consider the enormous power of 'the meta' by thinking about this extremely counter-intuitive assertion:

> At the metalevel, any two contradictories are one thing, because they choose the same distinction from the void.

I should argue that oriental wisdom is based quite fundamentally on this assertion, whereas Western wisdom has put the reductionism of polar categorisation into high gear to do its donkey-work. That has resulted in an 'advanced' civilization, which unfortunately carries with it all the disadvantages of the dichotomy trap. No-one who understood my assertion would want to burn his opponent at the stake. Anyone who wishes to consider this way of looking at the world more carefully should study oriental philosophy, and also make sure to read Spencer Brown[3] who invented a mathematics of great power and beauty precisely to handle these matters.

Let us turn to the practical consequences. This story is perfectly practical, though maybe apocryphal. A little girl surveyed my long white beard for a time and then asked: 'Were you in Noah's Ark?' I replied that I was not. 'Then why weren't you drowned? Notice the precise point. These are polar opposites in a universe of discourse that happens to be irrelevant. The irrelevance can be detected only metalinguistically. When a head of state justifies having done something irretrievably silly, it is usually because his categorization of the problem did not permit it to be newly defined. When another head of government declares that 'there is no alternative', it is usually because the structure of her logic vouchsafes none. To redefine problems and to redesign systems are our only recourses when things manifestly do not work. To do either of these entails detecting a metasystem and learning the metalanguage that it speaks.

The Readings printed in this book address these issues from various points of view in various styles. Professor van Gigch introduces them with an 'Overview'- which necessarily speaks the language of the book. This metacomment does not need to do that. I want to declare my belief that the problems here addressed are at once real and important; that the 'method of meta' for their resolution is not only valid but the uniquely possible approach; that the practical consequences of getting such ways of thinking adopted would be little short of sensational; and that therefore the book is both important and timely. In this way I exemplify the advocacy of the book: it is not necessary, it is not even relevant, for me to discuss the Readings or to take up the occasional disagreement that I might have with an author. Because this is a *meta*comment: it is a comment on the book, and not on the contents of the book.

In this example we have indeed an exemplar for the 'method of meta'. We should be looking at problems themselves, and not at their content. The likelihood is that the content of the problem has been generated at root by limitations of the brain as a machine for reasoning, and by the limitations of the logic which that reasoning-machine is able to create. But the problem as such is a pathological symptom of a societary system. The discussion of that pathology is metalinguistic to the internal language of the problem, which everyone is avidly discussing in parliament, in the media, and in pubs alike.

Consider arguments about pay and conditions between unions and management. All participants, all commentators, burrow about in the details of the problem: they examine the content. But the social pathology has to do with the distribution of wealth, and the adaptation of whole neighbourhoods and brotherhoods. These issues are metasystemic to the content of specific problems, where output is lost, where strikes occur, and arbitration speaks the problem-language by rote. The trouble is that metalinguistic discussion twenty years ahead of the wholly predictable problem is usually regarded as likely to precipitate that problem. Responsible people on all sides hope that the wholly predictable problem will go away, not happen at all or at least will not happen until they personally have retired.

I am not theorising here. I was personally involved in these affairs in steel, in cotton, in shipbuilding. Outstandingly, I was involved (as a senior manager) in printing. Twenty years ago I was having amicable discussions and making practical progress with the British printing unions about the inevitable automation of the industry. These discussions were metalinguistic: what is a safe and sure process to ensure adaptation in the industry under a changing environment, whereby the livelihood of everyone and the profits of owners could both be secured? But management (and not the unions) would not grapple with the metasystemic problem because they saw it as provocative, and the systemic problems as such were more-or-less under control (at exorbitant cost). This week, as I write, I have watched newsreels of policemen charging print-union demonstrators on horseback and beating them with truncheons. The problem was allowed to polarize at the systemic level, whereas the meta-problem had shown itself to be tractable.

Here is the next sample problem that is urgent today. The big nation states exert enormous economic 'clout' and highly threatening military power over small nation states. In cybernetic terms, these are variety amplifiers of high gain: the variety by which the small nation can express its identity at home and in the foreign councils of the world is overwhelmed and absorbed by the variety which counters its every expression with variety overkill. What can the small nation state do? It has to try and restore the equation of requisite variety by some method of variety amplification of its own that cannot be absorbed and nullified by the variety blanket of the super-power.

The name of this answer is terrorism. It is another wholly predictable problem. It is manifestly impossible for a surveillance system to generate protective requisite variety in face of the assassin and the aircraft hijacker, because it cannot discriminate between all possible states of the system. Therefore it is necessary to handle the issue at the metalevel, whereby the initial super-power stance is shown to be generating the succession of actual problems that result in so much havoc. It further follows that to increase the variety amplifications of the super-power yet further by armed assault on the small nation is necessarily counter-productive. It must produce escalation, another wholly predictable problem lying ahead. As usual, we need to redefine the issues at the metalevel, and to redesign the system that generates predictable problems.

My final example shall concern the nation states themselves, and the aspiration to world governance enshrined in the United Nations Organization, the Declaration of Human Rights, and so on. Now the network of sovereign states is a multi-dimensional set of polarized relationships. They confront each other in pairs, dichotomizing issues between them; they confront each other in groupings of various sorts, dichotomizing again as in East/West and North/South, not to mention rich/poor. We need to remember that most of these nation states exist in conditions of great poverty and therefore malnutrition. At least half live under oppressive dictatorships, where human rights are largely disregarded. Others live under 'elective dictatorships' where democracy is warped and inept. Suppose we assume that most people on the planet would like to live in peace and contentment, fed and sheltered, in collaboration and with concern for others. It is evident that the network of sovereign states cannot give expression to these hopes.

Rulers are paranoid and their administrations are corrupt in all their various ways: it is not systemically possible that any nation state will voluntarily yield any degree of sovereignty to another, or to the total system because that can be manifested only as a subset of its own parts. The stream of wholly predictable problems that flows out of this inoperative system is with us all the time. As usual, the world is condemned to discuss them *in the problem language*, which reifies and entrenches the dichotomies that the problem is all about: armament/ disarmament at the head of a very long list.

Is anyone speaking the metalanguage that makes all of this exceedingly dangerous absurdity an irrelevance? Of course. They are the majority of planetary citizens whom we supposed wish to live in peace and contentment. Their inarticulate aspirations are metalinguistic to the language of sovereign states, and they themselves *are* the metasystem. This is not the place to try and express the consequences that ought to follow from this example. However we should note the existence of a movement which tries to embody the metasystem that already exists.[4] This is surely the insight that metasystems, being logical constructs, already exist. This is what matters in this as in the earlier examples.The task is not to preach Utopia, not to hanker after reforms in existing systems that will always be blocked (if they really matter) with all the skill, resources and determination that the status quo can muster. Rather it is to give expressions to metasystems that render the problem-generating systems irrelevant. There will be new problems, no doubt, generated at *that* level of systemic design. Ah! But metasystems are recursive . . . . we do not have to be trapped by our own logics unless we enjoy it, and to know that is the beginning of spiritual progress.

We have technologies now that are capable of linking all the world's peoples in a network, which I have called the Technosphere following on the Geosphere and Biosphere (and before reaching Teilhard's Noosphere). The term is meant to convey the pervasiveness, rather than the point-to-point nature, of communications. So the metasystem even has its own technology to hand. What is left is the ability of folk to see the point. It can happen only by the laborious process of rethinking, of changing paradigms.

May I commend this book to its readers for its endeavour to work through the 'method of meta' in the various ways of its many authors, thereby provoking new insights and challenging stale beliefs. So far our practice has been to try and solve a societal problem within the system that generated it, using the language that defines the system to be as it is – and not otherwise. The solutions therefore have one thing in common. They do not work.

REFERENCES

1. McCulloch, Warren S., *Embodiments of Mind*, MIT Press, 1965.
2. Beer, Stafford, discussion of 'Two Theorems of Statistical Separability in the Perceptron' *Mechanisation of Thought Processes*, National Physical Laboratory, Her Majesty's Stationery Office, 1959.
3. Spencer Brown, G., *Laws of Form*, George Allen and Unwin, 1969.
4. World Service Authority 1012 14th Street N.W. Washington D.C. 20005 U.S.A.

ABOUT THE AUTHORS

John R. Battista (*Informational Holism: Toward an Integration of the Holographic and Analytical Models of Consciousness*, Chapter 10), is a psychiatrist who has served as the Clinical Director of The Country Place, a residential treatment program in Litchfield, Connecticut. He teaches in the Psychiatry Departments of Yale University and the University of Connecticut, as well as serving on the faculty of Temenos Institute. His research interests involve the use of a systems framework to develop an integrating theory of personality and the nature of consciousness. For further information see Battista, J.R., 'Self Psychology and Systems Theory', in *General Systems Theory and the Psychological Sciences*, edited by W. Gray, J. Fidler, and J.R. Battista (Oceanside, California: Intersystems Press, 1983). *Address/Affiliation:* Rd 1, Cherniske Rd, New Milford, Connecticut, 06776, USA.

Stafford Beer ("Metacomment", Foreword, and "The Argument for a Metasystem Based on Stafford Beer's Thinking", Chapter 1). Beer's spirit is not only present in his "Metacomment" but is also evident throughout the rest of the text. Without a doubt he is the most illustrious and most prolific of the authors who contributed to this work. He remains the foremost cybernetician of our time and, as such, advises private corporations, industries and governments through the world. He has received countless prizes and awards for his work in Cybernetics from British, Swedish and other world organizations. His most important books are *Cybernetics and Management* (1959), *Decision and Control* (1966), *Brain of the Firm* (1972, 1981), *Designing Freedom* (1974), *Platform for Change* (1975), *The Heart of Enterprise* (1979), *Diagnosing the System for Organizations* (1985) He has even ventured into the world of poetry with *Transit* (1977, 1983). *Address*: Cwarel Isaf, Pont Creuddyn, Lampeter, Dyfed, Wales SA48 8PG, U.K.

N. A. Coulter, Jr. (*Teleogenic Systems Theory*, Chapter 12), is Professor of Surgery and Physiology at the University of North Carolina School of Medicine, and Director of the Biomedical Engineering and Mathematics curriculum. His research interests are in teleogenic systems theory, artificial intelligence and synergetics. He is the author of *Synergetics: An Adventure in Human Development. Address/Affiliation:* Biomedical Engineering and Mathematics Curriculum, Dept. of Surgery, School of Medicine and School of Social Work, University of North Carolina, Chapel Hill, North Carolina, 27514, U.S.A.

Nora Femenia (*The Metasystem Approach and the Dysfunctional Family System*, Chapter 21), is a Professor of Clinical Psychology at the Universidad de Belgrano and at the Institute of Cybernetics, both in Buenos Aires, Argentina. Her academic training is in psychology, psychiatry and psychotherapy, and she has established a successful private practice in these fields. Her current consulting interest and publications are concerned with the application of systems and cybernetics concepts to the fields of her specialties. *Address/Affiliation:* Asociación Argentina de Teoría General de Sistemas y Cibernética, Carabelas 281, 7to., "H", 1009 Buenos Aires, Argentina.

C. O. François (*Consequences of Gödel's Theorem for the Metasystem Paradigm*, Chapter 5) combines the career of a diplomat with that of a systems scientist. He resides in Buenos Aires, Argentina, where he has helped to create a study group which aims to disseminate systems ideas in the community. François has published in French and in Spanish. He is the author of *Phénoménologie, Cybernétique et Prospective: Un Etude des Fondements de la Prospective* which has been translated into Spanish. He is also on the editorial boards of important international journals and has sparked many innovative projects in the systems sciences. *Address/Affiliation:* Asociación Argentina de Teoría General de Sistemas y Cibernética, Carabelas 281, 7to., "H", 1009 Buenos Aires, Argentina.

Ann F. van Gigch (*The Application of the Metasystem Paradigm to Cost Containment in a Health Care Program*, Chapter 20) is a Lecturer in Management in the Department of Management, School of Business and Public Administration, California State University, Sacramento. Given her nursing background and her education in Business Administration, she is well equipped to deal with the new developments in health care which loom on the horizon. She is a Stafford Beer admirer and enjoys the application of systems thinking to solve the problems of complexity which plague the health care industry. *Address/Affiliation:* Department of Management, School of Business and Public Administration, California State University, Sacramento, California, 95819, USA.

J. P. van Gigch (*Editor of the text, author of the Introduction and Overview*, Chapters 1, 2, 4, 18 and 19 *and co-author*, Chapter 3) is a Professor of Management. He specializes in Systems Theory and Research Methodology. He is the author of *Applied General Systems Theory* (Harper and Row, 2nd edition, 1978), which has been translated into Russian and Spanish. The Chinese translation is also in process. He has published widely in several international journals and serves on the Editorial Board of *Human Systems Management* (Amsterdam, North Holland). His research objective is to develop methodologies which are appropriate to the solution of the problems of complexity of social science domains. *Address/Affiliation:* Department of Management, School of Business and Public Administration, California State University, Sacramento, California, 95819, USA.

Ranulph Glanville (*Very Rarely Stable*, Chapter 16) is Senior Lecturer in Architecture and Cybernetics at Portsmouth Polytechnic, Visiting Professor of Andragology at the University of Amsterdam, and Special Project Organizer at the Architectural Association, London. After initial training as an architect, he took his PhD in Cybernetics under Gordon Pask at Brunel University, being examined by Heinz von Foerster. As well as teaching, he researches spatial perception, design theory, design education, and abstract cybernetics. He lectures extensively and has published about 80 works. He is European Editor of *Cybernetics Forum*. His most recent paper is 'Theory of Model Dimensions Applied to the Computer Solution of a Syllogism', *International Journal Man–Machine Studies*, Summer 1983. *Address/Affiliation:* Portsmouth Polytechnic School of Architecture, King Henry I Street, Portsmouth, Hants, PO1 2DY, UK.

A. L. Johnson (*Teleogenic Systems Theory*, Chapter 12) is an Associate Professor in the School of Business and Social Work at the University of North Carolina, Chapel Hill. His research interest is in organization theory, management of service organizations, and health policy. *Address/Affiliation:* School of Social Work, University of North Carolina, 27541, USA.

Walter J. M. Kickert (*A Metasystem Approach to Organization Decision Making*, Chapter 3) graduated in Experimental Physics, and after working at London University in fuzzy set theory and control theory, wrote a dissertation on Organization Science. Since then, he has worked at the Public Administration group of the Institute for Political Science of Nijmegen University in The Netherlands. His current research interests are organization, policy making and planning in public administration, systems theory, and particularly, flexibility and cutbacks in public administration. Recent publications are *Organization of Decision Making* (North Holland, Amsterdam, 1980), and several Dutch articles on planning and flexibility. His contribution to the metasystem paradigm should not be underestimated. He sparked the seminal paper on the subject. *Address/Affiliation:* Institute of Political Sciences, Katholieke Universiteit, 6500 HK Nijmegen, The Netherlands.

Vladimir A. Lefebvre (*Metasystems and Problems of Freedom of Choice: New Arguments in Favor of an Old Metaphor*, Chapter 9) was born in the USSR and received his doctoral degree at Moscow State University. His book, *Conflicting Structures* (Moscow, 1967), was the first Soviet book in the field. It was later called *Cybernetics of the Second Order*. This book has significantly influenced the Soviet paradigm in cybernetics and cognitive science. It was published in the USA under the title *The Structure of Awareness* (SAGE, 1977), translated by Anatol Rapoport. In his recent book, *Algebra of Conscience* (D. Reidel, 1982), Lefebvre describes a mathematical model of ethical cognition which establishes some formal connections between an individual's inner world and his behaviour. *Address/Affiliation:* School of Social Sciences, University of California, Irvine, California, 92717, USA.

Magoroh Maruyama (*Communication Between Mindscapes Types*, Chapter 6) was born in Japan, studied at Universities of California (Berkeley), Munich, Heidelberg, Copenhagen, and Lund. Received his BA in Mathematics and PhD in Cultural Epistemology. He has been on the faculty at Stanford University, University of Illinois, Urbana, and University of Uppsala; worked in Asia, Europe, and the Middle East, and has been consultant to numerous multi-national corporations and government agencies. He also taught International Business at the University of California, Los Angeles. *Address/Affiliation:* Department of Business Economics, University of Hawaii, 2404 Maile Way, Honolulu, Hawaii, 96822, USA.

Richard O. Mason (*Dialectical Pragmatism: A Progress Report on an Interdisciplinary Program of Research on Dialectical Inquiring Systems*, Chapter 11) is Professor of Management and Management Information Systems, School of Business and Public Administration, University of Arizona. He received his PhD from the

University of California, Berkeley (1968). Mason has been involved in research and consulting in the fields of information systems, strategy and organizational behaviour with a number of public, private and not-for-profit organizations. His current interests include organizational strategy, the behavioural side of information systems and the ethics of information. He is a co-author of *The 1980 Census: Policymaking Amid Turbulence*, *Challenging Strategic Planning Assumptions*, *Creating a Dialectical Social Science*, *Measurement for Management Decision*, and *Strategic Management and Business Policy* as well as other books and publications. *Address/Affiliation:* Carr P. Collins Distinguished Professor of MIS, Edwin L. Cox School of Business, Southern Methodist University, Dallas, Texas, 75275, USA.

Ian I. Mitroff (*Dialectical Pragmatism: A Progress Report on an Interdisciplinary Program of Research on Dialectical Inquiring Systems*, Chapter 11) is the Harold Quinton Distinguished Professor of Business Policy in the Department of Management and Policy Sciences, University of Southern California, Los Angeles, California. He received a BS in Engineering Physics, a MS in Structural Mechanics, and a PhD in Engineering Science and the Philosophy of Social Science all from the University of California, Berkeley. Mitroff is a member of the American Association for the Advancement of Science, Academy of Management, American Psychological Association, American Sociological Association, Philosophy of Science Association, and The Institute of Management Science. He has published over 130 papers and six books in the areas of business policy, managerial psychology, strategic planning, and the philosophy and sociology of science. *Address/Affiliation:* Graduate School of Business, University of Southern California, Los Angeles, California, 90007, USA.

Maria Nowakowska (*The Dynamics of Knowledge Structure: Semiotic Systems*, Chapter 8) is Associate Professor and head of the Laboratory of Psychology and Action Theory in the Institute of Philosophy and Sociology, Polish Academy of Sciences, Warsaw. She is the author of six books (with another three in English, in press). She has contributed over 30 chapters in books and written more than 100 articles in professional journals. She feels that her most important work deals with the formal theory of actions (*Language of Motivation and Language of Actions*, Mouton, 1973). This book received the Prize of the Scientific Secretary of the Polish Academy of Sciences. She also writes in the areas of mathematical psychology, decision theory, mathematical sociology, linguistics, semiotics, systems theory, and philosophy of science. *Address/Affiliation:* Institute of Philosophy and Sociology, Polish Academy of Sciences, Warsaw, and Machine Intelligence Institute, Iona College, New Rochelle, N.Y., 10801, USA.

R. K. Ragade (*Metagames and Metasystems*, Chapter 13 and Concluding Remarks) is a Professor of Engineering Mathematics and Computer Science, University of Louisville, Kentucky. While not participating actively in the editorial tasks of this text, Dr Ragade was instrumental in providing important ideas for it and has been an important force behind the whole project. His current research interests are in Quantitative General Systems Theories and their applications, Systems Modelling and Simulation, Information Systems Design, Organizational Cybernetics, Office Automation, Artificial Intelligence, Fuzzy Set Theory and Systems Philosophy, Analysis of Systems Formation in Multipurpose Systems. It is in the latter, Dr Ragade claims, that the metasystem paradigm offers most potential. Dr Ragade has been Editor of *General Systems*, the Yearbook of the Society for General Systems Research, since 1980. *Address/Affiliation:* Speed Scientific School, University of Louisville, Louisville, Kentucky, 40292, USA.

Mildred L. G. Shaw (*Epistemological Hierarchies in Knowledge and Expert Systems*, Chapter 14) has pioneered the development of techniques for computer-based elicitation of personal construct systems. She has authored *On Becoming a Personal Scientist* (London: Academic Press, 1980) and *Recent Advances in Personal Construct Technology* (London: Academic Press, 1981) as well as an innumerable number of articles, monographs, and computer programs where the methodology for knowledge acquisition is discussed. Her work places her at the forefront of development of 'user-friendly' expert systems. *Address/Affiliation:* Department of Computer Science, University of Calgary, Alberta, T2N 1N4, Canada.

Werner Ulrich (*The Metaphysics of Social Design*, Chapter 17), born 1948 in Berne, Switzerland, is presently Head of the Office for Evaluation Research at the State Department of Public Health and Social Services, Berne, Switzerland. He holds a Dr.Rer.Pol. from the University of Freiburg, Switzerland, and a PhD in Philosophy of Social Systems Design from the University of California at Berkeley. His research interests comprise the philosophical foundations of applied science, especially of management and planning theory: heuristics, especially the idea of a critical systems approach to planning; practical philosophy; and evaluation research. Among over 30 publications on these subjects, his most important one thus far is *Critical Heuristics of Social Planning. A New Approach to Practical Philosophy* (Berne, Switzerland: Haupt 1983), a comprehensive study on the epistemological and heuristic foundations of a critical systems approach to planning and systems design. *Address/Affiliation:* Office for Evaluation Research, State Department of Public Health and Social Services, Rathausgasse 1, CH–3011 Berne, Switzerland.

Burton H. Voorhees (*Systems, Models, Methods: Epistemic Instrumentation*, Chapter 7) was educated at the University of California, Berkeley (1960–64), The University of Arizona (1964–66) and the University of Texas at Austin (1966–71) where he received his PhD in physics, specializing in general relativity. His major current interest is in human mental processes, including cognitive/affective and sensory/motor processes, creativity and socio-cultural historical processes. He is interested in modelling such processes, and in teaching methods which further the development of human potentials. He is a member of the Society for General Relativity and Gravitation, the Society for General Systems Research and Arica Institute. His most recent publications have appeared in the *Bulletin of Mathematical Biology*, *Behavioral Science* and the *Int. J. of Theoretical Physics*. *Address/Affiliation:* Athabasca University, 12352–149th St., Edmonton, Alberta, T5V 1G9, Canada.

Gerard de Zeeuw (*Replacing Models by Actors: Some Ideas for Metadesign*, Chapter 15) studied pure mathematics, mathematical statistics, econometrics, and mathematical psychology at the Universities of Leyden, Rotterdam, and Stanford. In his PhD thesis he studied various methodological problems of 'model-thinking in psychology'. After working for nine years as a researcher and as a consultant in psychological research, he became Professor of Social Change, especially of its Research Methodology (the discipline is called Andragology in The Netherlands). His research interests vary over research in education, in artificial intelligence, in systems theory, in social aspects of computerization, and science of science. He has organized more than 35 scientific meetings and conferences and has given invited talks in many universities. He is author or editor of six books, and has written more than 200 papers. *Address/Affiliation:* Subfaculty for Andragology, University of Amsterdam, Grote Bickersstraat 72, 1013 KS, Amsterdam, The Netherlands.

SECTION I

INTRODUCTION

DEFINITIONS

John P. van Gigch

METAMODEL A metamodel is a model of a higher level of abstraction which maps the properties of lower systems models into propositions of a higher level of abstraction. Of necessity, the metamodel is conceptualised in a metasystemic language or metalogic which can decide propositions and reconcile arguments which cannot be decided or reconciled in the language or logic of lower systems models.

(After Stafford Beer)

METAMODELLING Until recently, the accepted approach to problem-solving consisted of the formulation of a model and applying the solution obtained from the model to the real world. However, this methodology bypasses the important step of metamodelling, that is setting the rules of logic by which the model itself is to be designed. Metamodelling, or the Science of Modelling, is involved with designing metamodels and with designing the process of design.

METASYSTEM A system over and above a system of lower logical order, and therefore capable of deciding propositions, discussing criteria, or exercising regulation for systems that are themselves logically incapable of such decisions and discussions, or of self-regulation (because the metalogic is inaccessible to the system's logic, or the metalanguage is capable of statements inexpressible in the system's language).

(After Stafford Beer)

METASYSTEM PARADIGM As a way of thinking, the metasystem paradigm is a form of metamodelling which is used to conceptualize and study decision-making in complex hierarchical systems at several levels of recursion.

METALANGUAGE A language of a higher order of logic which is spoken at the metasystemic level. Metalanguages are needed to resolve debates, conflicts or undecidability which cannot be resolved in the language or logical structure of systems at the object level.

(After Stafford Beer)

OVERVIEW

John P. van Gigch

The metasystem paradigm arose from the need to establish a methodology by which conflicts at lower levels of logic could be resolved. Neither the traditional scientific method or the more contemporary systems approach have explicity taken into account the epistemological problems which arise from the establishment of organizational hierarchies.

In Chapter 1 (*The Argument for a Metasystem Based on Stafford Beer's Thinking*) van Gigch summarizes the argument in favour of the metasystem concept, as elaborated by Stafford Beer in his Metacomment (**Foreword**) and in his many other writings. Beer, more than anyone, has pleaded for the institutionalization of the metasystem, lest our society fall apart. Beer foresaw the need to formalise the existence of the metasystem in the organizational context, a system over and beyond a system of lower logical order, capable of deciding propositions, discussing criteria, or exercising regulation over systems that are themselves logically incapable of such decisions or of self-regulation. Excerpts from *Platform for Change* illustrate Beer's witty description of the role of a metasystem, metalogic, metalanguage and metamodels.

The concepts of metasystem, metadesign, metamodelling are not new. Under other names, they have been treated in innumerable contexts. Thus, metaphysics and meta-ethics have been subsumed under the subject of philosophy and logic. What is novel, is to study the role of the metasystem and to develop its implications in the context of organizational decision-making.

In Chapter 2 (*Methodological Comparison of the Science Systems and Metasystem Paradigms*), van Gigch compares three paradigms (science, systems and metasystem paradigms) to situate the reader within the methodological framework of contemporary approaches used by the scientific community to elaborate solutions to the vexing problems of complexity. The domain of each of the paradigms as well as the nature of the solutions obtained, the criteria to evaluate and guarantee truth are highlighted. The shortcomings of the science and the systems paradigms thus anticipate the need for the metasystem approach which the present text attempts, in a modest way, to satisfy.

Starting from the simple control system (comprising controller and controlled system) and its metacontroller, Kickert and van Gigch (Chapter 3, *A Metasystem Approach to Organizational Decision-Making*) develop the metasystem approach as a framework of decision-making about decision-making, that is, the structuring of the decision-making process and the determination of organizational objectives. Like Beer, these authors call for the explicit incorporation of metadecision-making models in organizational design to set out the roles of system and metasystem.

When a paradigm shift is contemplated, it is pertinent to review whether the new paradigm is liable to answer epistemological questions left unanswered by the old one. Van Gigch proceeds to recall these questions in Chapter 4 (*Epistemological Questions Raised by the Metasystem Paradigm*).

François (Chapter 5, *Consequences of Gödel's Theorem for the Metasystem Paradigm*) contests the claim that the metasystem paradigm may ever be the answer to the search for a guarantor of truth, given Gödel's Theorem which asserts that all generally used logical systems are incomplete, if they are consistent. There always exist true statements expressible in the language of these systems which cannot be proven in the logic of the system itself. One must always go outside a given system to establish a criterion of its validity.

In the rest of the book, contributors do not always refer specifically to the metasystem paradigm. However, it is clear that each one offers a metasystemic view. Epistemology, philosophy of science and discussion about the art or science of modelling can only be carried out at the metalevel, in the language of the metasystem (a metalanguage and/or metalogic).

Maruyama (Chapter 6, *Communication Between Mindscape Types*) explains how theories, logics and epistemologies constitute three distinct levels. Different theories can be built on the same logic, different logics may stem from the same epistemology. He then proceeds to illustrate the characteristics of four epistemologies or mindscapes. He discusses how lack of natural understanding among persons with different mindscapes arise and speculates that the western world is ripe for an alternative logic. He calls for two types of wisdom: that which exists in specific cultures and disciplines and that which is gained from transepistemological understanding.

Voorhees (Chapter 7, *Systems Models, Methods: Epistemic Instrumentation*) searches for a new model of science. He posits a scientific paradigm due to Boyce called psychological epistemology, in which three independent ways of knowing called rationalism, empiricism and metaphorism exist. The introduction of metaphorism is said to augment the rational-empirical epistemology where systems are considered a bridge between empiricism and metaphorism. In this scheme, scientific knowledge begins as metaphor and becomes paradigm with the choice of a unique set of models for the systems and its components.

Theory of formal semiotics (*The Theory of Local Systems*) and knowledge representation are discussed by Nowakowska (Chapter 8, *The Dynamics of Knowledge Structure: Semiotic Systems*). The value of formal semiotics lies in the possibility of unification of several approaches in cognitive processes, raising crucial problems about the structure of common knowledge and belief. She illustrates an application of formal semiotics to representation of knowledge and argues that the semiotical approach offers clear and well established logical research on the structure of knowledge and its relation to reality. For philosophy, it offers a unification of ontology and epistemology where the classical scope of the latter is extended. For logic, logical calculi are simplified and improved. For linguistics, it allows the consideration of a unified system lying at the base of all natural and formal languages.

Lefebvre (Chapter 9, *Metasystems and Problems of Freedom of Choice: New Arguments in Favor of an Old Metaphor*) tackles the topic of analogies between the conscious actions of man and the behaviour of elementary particles or electrons. He analyses human inner feelings in a model of ethical cognition where the human inner world is represented by a tree-like hierarchy of meta-images. He draws an ingenious analogy between the model of quantum mechanics, illustrating the interference of elementary particles through two parallel slots onto a screen, and his algebraic models of ethical cognition. He demonstrates that it is impossible to describe functionally how an individual decides who is responsible for a negative situation, if this individual does not have correct registrations of his inner images. If such a choice is called freedom of will, then, according to Lefebvre, the concepts of freedom of will and of indeterminism are equivalent.

In a similar view, although avoiding Lefebvre's world of mathematical phychology, Battista (Chapter 10, *Informational Holism: Toward an Integration of the Holographic and Analytical Models of Consciousness*) attempts to draw a general theory of consciousness by providing a general structure which relates different forms of consciousness. His theory rests on three basic concepts: (1) consciousness is information; (2) the different forms of consciousness refer to different hierarchical levels of information and (3) the intensity of consciousness at any level, is a function of the amount of information at that level. In his theory of informational holism each state of consciousness has a metalevel which, at the highest level, results in pure awareness or an integrated awareness of all other levels. This theory is said to clarify the relationship between brain and consciousness and even to make predictions about the meaning of consciousness in non-human systems.

As explained by van Gigch in Chapters 18 and 19, inquiring systems are metasystems whose epistemologies provide rationality to object systems. Mitroff and Mason (Chapter 11, *Dialectical Pragmatism: a Progress Report on an Interdisciplinary Program of Research on Dialectical Inquiring Systems*) discuss a particular inquiring systems – the dialectical inquirer. They attempt to create a dialectical social science which can be subjected to empirical scrutiny, which can accept both qualitative and quantitative models of the dialectic, and where the philosophic dialectic concepts can be operationalized as a methodological tool. Thus, the dialectic is offered as a fruitful model of organization, as a communication theory, as a theory of experimentation, as a training experiment, as a generalized theory of planning, as a process of reason and, finally, as a general theory of social science.

Teleogenic Systems Theory (Coulter and Johnson, Chapter 12) deals with systems that are capable of generating their own goals. These authors explore how teleogenic systems can contribute to the development of a metasystem methodology. They describe the field and domain over which a teleogenic system applies, the problems to which the methodology is suited and the nature of the solutions obtained. They discuss how truth in teleogenic systems can be evaluated and whether it can be guaranteed. Thus, this approach can easily be compared to other methodologies such as those outlined in Chapter 2 by van Gigch. Ideas from this chapter can also be used as tentative answers to some of the epistemological questions raised by van Gigch in Chapter 4.

In Chapter 13 (*Metagames and Metasystems*) Ragade considers the introduction of various possible cognitive processes in an interactive strategic decision analysis environment. It is thus related to Shaw's paper (Chapter 14) on expert systems that follows. A metagame is used to model a player's particular cognitive process about the other player's cognitive processes, such that the outcomes in the game are seen as reasonable to him/her. A metagame is a game derived from a basic game. It is a metalogical extension, through which bounds of rationality are seen reduced in the basic game. Rational, symmetrical metarational and general metarational outcomes are described and explained.

In a more concrete vein, Shaw (Chapter 14, *Epistemological Hierarchies in Knowledge and Expert Systems*) discusses the development of expert systems in which the computer acts as a medium for access to encoded expert knowledge. First, she shows the importance to establish a terminological framework in terms of Klir's epistemological hierarchy. She then relates it to Kelly's work on personal construct systems, to Shaw's techniques for computer-based elicitation of constructs and to Zadeh's Theory of Fuzzy Sets. She finally demonstrates the value of this approach by using a computer program to elicit a basic vocabulary and knowledge structures in a particular problem area.

De Zeeuw's and Glanville's contributions (Chapter 15, *Replacing Models by Actors. Some Ideas for Metadesign* and Chapter 16, *Very Rarely Stable*, respectively) should be read as a unit. De Zeeuw shows how the introduction of a solution to a particular problem brings about the advent of a new context where the original solution is no longer dominant. Thus, with the new context, new problems arise called post-solution or PS-problems. De Zeeuw uses the word 'context' to designate a new domain of solutions. As we solve the problem, the context (or domain) changes, bringing about the necessity to consider a new set of possible problems and, as a result, of solutions. De Zeeuw reminds us of the relation between the one-actor and the two-actor models and the differentiations between information 'models for' and 'models of' the metaworlds.

In *Very Rarely Stable* (Chapter 16), Glanville continues de Zeeuw's theme and dwells on the 'model for', as a dynamic approach to think about and solve problems. Glanville considers methodology, the method of method that is, a metamethod. With his tacit permission, we would ascribe him, in Maruyama's terminology (Chapter 6), a morphogenetic epistemology. Indeed, Glanville's refusal to believe in levels and boundaries, his turn of mind which consists of using drawings to explore and develop ideas, where the drawing is a model of an idea while it also affects the idea – this reflexive and circular design approach – draws him ever so close to Maruyama's G-mindscape.

Ulrich (Chapter 17, *The Metaphysics of Social Systems Design*) joins Glanville and Maruyama in elaborating metaprinciples of social systems design. In a fictitious dialectical debate Ulrich opposes the views of two philosophers and systems scientists. He criticises theories of complexity which 'suffer from a common inability to deal critically with the very social reality which they ought to improve' and leans in favour of an approach very similar to that elaborated by Glanville above, where the planner/designer is not provided with analytical tools, but rather dialectic tools, to help him reflect on his designs and enter into a reasonable discourse with those who will be affected by them.

In Chapter 18 (*The Metasystem, Rationalities and Information*), van Gigch elaborates the metasystem paradigm into a practical methodology to diagnose organizational dysfunctions. He uses the control system framework elaborated earlier by Kickert and van Gigch (Chapter 2) to show how four kinds of rationalities are used for decision-making. The differentiation between data, information and intelligence is drawn to explain where and why decision-making leads to systems failures.

Chapter 19 (van Gigch, *Rationalities and Metarationalities in Organizational Decision-making*), is an extension of Chapter 18. The metasystem paradigm is applied to organizational decision-making which is regarded as a multilevel process taking place in an inquiring system which converts inputs (evidence) into output (decisions), and whose epistemology consists of rationalities and metarationalities. In this chapter, an attempt is made to resolve the question of problem isomorphy by studying how the universe of discourse changes from level to level in several illustrative examples.

Chapter 20 (Ann van Gigch, *The Application of the Metasystem Paradigm to Cost Containment in a Health Care Program*), illustrates the application of the above methodology to the case of the health care system whose costs are out of control. It should be read carefully by managers and planners who seek a real-world approach to pinpoint and solve organizational deficiencies and malfunctions.

Chapter 21 (Nora Femenia, *The Metasystem Approach and the Dysfunctional Family System*) is another example of direct application of the metasystem paradigm to a social science domain, in this case, the family system. The metasystem paradigm proves to be an effective approach to study how the family system and the systems that lend it support in society, can be modified or reorganized to avoid present-day dysfunctions.

Ragade (*Concluding Remarks*), speculates how the ideas of this volume can be extended by further research. He asks two very pertinent questions. One, whether the metasystem paradigm is justifiable in view of the variety of perspectives which require the emergence of new systemic levels, and two, whether it would be more appropriate to call the metasystem paradigm a metaparadigm, that is, the grand paradigm of all systems paradigms by which all other systems paradigms are to be designed. Needless to say that answers to these questions are not forthcoming and are reserved for future endeavours.

1. THE ARGUMENT FOR A METASYSTEM BASED ON STAFFORD BEER'S THINKING

John P. van Gigch

The concept of metasystem is presented as conceptualized by Stafford Beer, one of the foremost cyberneticians of our time and to whom this book is dedicated. With his permission, excerpts from his poetry and prose are either paraphrased or reproduced herein.

We thought that a review of Stafford Beer's ideas would be a fitting introduction to a book of *Metamodels and Metasystems*. This distinguished cybernetician has already spelled out, in several of his books, the thesis and requirements for viable metasystems. The ideas of this introduction are, for the most part, gleaned from *Platform for Change*. We have consulted with the author who has agreed to the publication of this article containing his thoughts and several of his quotations.

As stated in the Foreword to the present text entitled *Metacomment*, Beer, more than anyone else in recent times, has pleaded for the "institutionalization" of the relevant metasystem "lest our society will fall apart". The summary of Beer's views will place the remaining contributions of this book in their appropriate context.

Beer conceptionalizes social systems and their environment as being made up of what he calls 'esoteric boxes'. An 'esoteric box' is not a closed system. It is self-organizing and self-regulating. It features complexity and unintelligibility to outsiders. The esoteric box represents a recognizable subsystem with recognizable boundaries. It is termed 'esoteric' because of the difficulty for outsiders or the uninitiated to know what goes on inside. Beer has emphasized the danger which managers face 'to mistake the model for reality and start managing the model instead'.

Beer characterizes the world we manage as being made up of 'strings and networks' of esoteric boxes, where the latter represent the subsystems of large systems. Usually, the subsystems pursue their own goals instead of contributing toward the larger system's purpose or mission. While on the one hand, subsystems are ultrastable and resist change and are dubbed 'machines designed for survival', on the other hand, the strings and networks of subsystems or esoteric boxes are unstable, 'mercurial' and their complexity defies the manager's attempts to apply rational decision-making. Beer calls these 'networks of complex connectivity': *metasystems*. This is 'the locale of society instability'. He wants the relevant metasystems to be identified and, in some sense, institutionalized. 'The purpose of a metasystemic social institution is to embody the *nerve centre* for metasystemic affairs. Its only function is, precisely, to identify situations of dangerous and therefore explosive instability, and to identify trajectories leading to stability.' Beer recommends the construction of *metamodels* whose solutions can be implemented in the metasystemic social institution.

WHY A METASYSTEM, A METALOGIC, A METALANGUAGE?

METALANGUAGE
a language
meta-μετα–
'the next after'
'over and beyond'
the language we are using at the time.

Why should anyone need a metalanguage?
There would be no point.
Unless the language we were using were defective.

All languages are defective.
In what special sense?
Simply that there are always propositions
about the language itself
that cannot be expressed *in* the language.

This discovery was made by theoretical logicians
(sometimes called metamathematicians)
and is generally regarded as 'academic'.
I do not think it is.
We can bring the idea down to earth:

> The barber in this town shaves every one
> who does not shave himself.
>
> Who shaves the barber?

It isn't that the barber has a beard.
The language-meaning the logical structure of the talking
is snarled up in itself.
There is no direct answer to the question.
More particularly, there is no way of discussing
what has gone wrong *using the language*
– because the language is in general perfectly all right.
You just find that you cannot talk about the barber himself
without contradicting yourself.

But we are already managing to discuss the problem
and we could soon sort it out.
Bertrand Russell did it first, and the key thought is about the class of classes
that are (or are not) members of themselves.
The key thought is nothing to do with either barbers or shaving.

Never mind the logical theory:
we are already speaking a metalanguage –
and I am sure you see why.

But everything we say –
and everything we even think –
is expressed in propositions having a logical structure.
If we have no formal way of questioning the *validity*
of saying any such thing as we say
we stand to be caught in this trap.

If the barber shaves himself, he cannot shave himself:
if he does not shave himself, he shaves himself.

The trap is very different from the thinking traps
of which we normally take account.
It has nothing to do with
    getting the facts wrong
    drawing false conclusions from the facts
    or taking too narrow a view

People are prepared for those traps.
And unfortunately they are inclined
to equate the problem we have raised
(which is called undecidability)
with taking too narrow a view.

This mistake, put in another form, says
a metalanguage offers a more embracing perspective
than the lower-order language.
It does do this – but the whole point is missed:

    a metalanguage is competent
    to discuss undecidability
    in the language

Just now we had a brief metalinguistic discussion
about the barber example, and spoke of
class inclusion.
Now we are in a position to construct
a new language for talking about this barber.

Look at these classes:

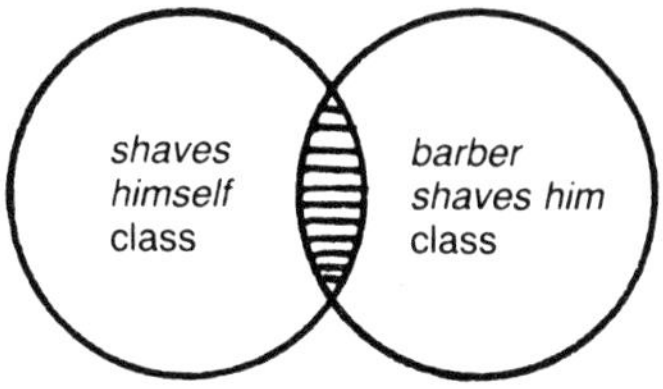

It is evident that the hatched area
IS the barber.
The language we started with is so constructed
that the two circles are *necessarily* kept apart.
That language actually DECREATES the barber.
No wonder we felt trapped.

Note that the second language is not a metalanguage
but that we had to move into a metalanguage
to provide the logical vantage point from which to percieve the nature of undecidability in the first language
and to design the second.

This understanding will be needed for further explanations later.
The purpose of explaining the term metalanguage now
in the context of the structure of this project and book
is to be able now to say:

> whatever is written in this explanatory colour
> is speaking the book's metalanguage.

Let us carefully consider the practical problem
presented by the project.

The only thing I need this metalanguage to discuss
is the emergence of the term –
which will be encountered frequently later
METASYSTEM.

Its meaning is really quite simple
providing that the original term
Metalanguage
was understood.

If a system speaks a language
appropriate to that system
the deficiencies of which
(its undecidable propositions)
can be discussed only in a metalanguage –

then the system that speaks the metalanguage

is a metasystem

There is no more to it than that.

The remark sounds apologetic –
thrown away
ingenuous . . . .

perhaps it is therefore disingenuous?

Not at all. And this is the big problem.
People seem to think that a metasystem
being 'over and beyond' the system to which it relates
has some special kind of authority –
that it secretly manipulates the system
and is sinister.

The power of a metasystem lies only
in its *logical* capacity to explain.

Beer proposes to institutionalize the metasystem to express and decide statements and arguments which cannot be expressed or decided at the object level. Only the metalanguage is competent to discern and settle undecidability existing in the language. One resorts to the metasystem to remove object level arguments from their context and concentrate on their logical structure. The metasystems can be instrumental in efforts to substitute stereotypes and the unfounded premise of conventional wisdom by value structures promoting higher order objectives which can lead to change and innovation.

In most on-going organizational structures, the control hierarchy is taken for granted and the triad of controlled system (CS), controller (CR), and metacontroller (MCR) are only identifiable after the fact. In other words, the existence of the metasystem or meta-metasystems is only happenstance but not the result of conscious design. Beer postulates the need to provide explicit structural rationality (Rs, in the notation of this text) by which metasystems for all subsystems and systems are built into the organization. In turn, the same metasystems can provide evaluative rationality (Re) which represents the interests and value structure of higher order systems.

The metasystem is needed to design an organizational structure capable of generating sufficient variety to absorb the variety exhibited by lower level systems, as well as to cope with the variety generated from the environment. One must design in the metasystem the appropriate variety generators, absorbers or reducers to match variety with variety.

It is interesting to note that metalanguages are parsimonious. They speak a higher level logic than object-level systems and can, usually, counter the latter's information with less transmission requirements. A metalanguage's influence does not reside in the authority or power stemming from its hierarchical position, but from its higher level logic. In Beer's scheme of things, metasystemic structures replace power structures. By doing so, lower level systems follow higher level objectives and, thus, optimize instead of suboptimize.

The interests of higher order systems will only be served if their values originate in the metasystem. To exercise control, the metasystem generates standards against which the output of lower level systems can be compared. The standards for control cannot originate in the same system in which the output is produced. Hence the need for the metasystem which can alter the internal workings of the ultrastable 'esoteric boxes'. This ultrastability and resistance to change can only be affected by imposing a higher order value structure originating in the metasystem. Furthermore, the metasystem provides the elements required for the strings and networks of esoteric boxes to be regulated and controlled to achieve a higher order synthesis.

Beer proposes 'eudemony' as a new metric, a new dimension by which the very metasystemic standards can be expressed. Beer flatly derides present norms which guide the esoteric boxes (subsystems) to produce misguided local objectives. As an example, he cites financial rewards as a poor metric to measure work (physical and mental). He declares that what is needed is a new value system. He calls for the design of eudemonic metasystems which allow decision-makers to make enlightened choices among alternative courses of action. 'They deal in indifferent choices of payoff rather than optional courses of action and these choices have to do with equilibria rather than the nature of the good.' The accepted and accustomed metrics will not do.

In his book *Platform for Change*, Beer proposes to devise metalanguages for any system whose problems cannot be decided in the language of the system. Here are some of the proposed metalanguages.

1. *A metalanguage of entropy* to rule over institutions in order to organize information flows which promote optimizations of metasystemic objectives instead of the suboptimization of lower systems goals.
2. *A metalanguage of complexity* to provide lower level systems with structural and evaluative rationalities originating in the metasystem to help resolve the mismatch between the intractable complexity of the real world and our perceived simplistic version of it.
3. *A metalanguage of stability* through which the calculus of variety absorption is

settled, and through which the difficulties of dealing with ultrastable esoteric boxes and the unstable metasystem are resolved.

4. *A metalanguage of boundaries* to replace and to counter the *metalanguage of categories.* The latter promotes the logic of synthesis and of holism while the former represents the logic of analysis and of reductionism.
5. *A metalanguage of conscience* to postulate value structures which can restore an individual's faith that attainable solutions to society's ills do exist.
6. *The metalanguage of coping* which provides humankind with the relevant ethic to allow them to survive 'the traps' set by the 'scale of their own perception', 'the dominance of money as the only measure of value, technology and a malignant bureaucracy'.

Finally, the metasystem paradigm may help in the resolution of apparent dilemmas and paradoxes inherent in the epistemology of systems science. The institutionalization of metasystems and of metalanguages provide a way by which dilemmas and choices between the opposite ends of spectra such as simplicity and complexity, optimization and sub-optimization, idealism and realism, incrementalism and holism, can be resolved.

In each case, the arguments appear unsolvable at the object level where the problem logic is of the 'go–no go' or (0,1) variety. However, when the dialogue is raised to a higher level of abstraction, the universe of discourse is not context-dependent but can be argued on the basis of logic and structure.

In many instances, these arguments revolve around epistemological questions such as the ones described in Chapter 4. The reader will note that Stafford Beer anticipated by about twenty years the questions raised here. In spite of our best efforts, we may, as yet, be lacking some of the answers.

2. METHODOLOGICAL COMPARISON OF THE SCIENCE, SYSTEMS AND METASYSTEM PARADIGMS*

John P. van Gigch

The science, systems and metasystem paradigms are compared on the basis of the field and domain to which they apply, the problems for which they are suited, the nature of solutions obtained and the criteria used for evaluating and validating the truth of their respective assertions.

INTRODUCTION

Several systems methodologies which exemplify three different paradigms are surveyed and compared. These methodologies illustrate the application of:

the science paradigm (section 3),
the systems paradigm (section 4) and
the metasystem paradigm (section 5).

Examples of methodologies which incorporate these paradigms are discussed under the following headings:

Description of Field
Nature of Domain
Problems to Which Suited
Application of the Paradigm
Nature of Solution (or 'truth' obtained)
Criteria to Evaluate 'Truth'
Proof and Guarantor of 'Truth'

For purposes of this study, *a paradigm* is defined as a distinct way of thinking about problems, in the sense given to this concept by Kuhn.[1] A paradigm is usually 'content' or 'substance-free' in the sense that it applies to many problems in a domain regardless of their specific content. A *methodology* is a problem orientated procedure or approach which incorporates a particular paradigm.

The intent is to show the characteristics of various methodologies presently available and to establish the requirements by which they may be evaluated.

* Reproduced with permission from: *Int. J. Man–Machine Studies* 11, 651–663 (1979). Copyright: Academic Press Inc. (London) Ltd.

LIMITATIONS

This paper refrains from being drawn into controversies regarding the origins of, or the reasons for, 'problemshifts' or paradigm changes. Also, the science paradigm as outlined here, presupposes the acceptance of inductive as well as deductive 'truth' without debating which is 'more scientific' in the sense of criticism levelled against 'subjective truth' by the logical reconstructionists.

Establishing definitions for what constitutes 'truth' and defining criteria for evaluating truth also constitute thorny philosophical questions which are beyond my intentions here. Within the limited confines of this paper, solutions obtained from problem-solving methodologies shall be labelled 'truths'. And it is assumed that to be valid, 'truths' should (1) meet criteria of internal coherence, (2) correspond as far as possible with perceived external reality, (3) be accepted by their recipients and (4) be morally justified. What constitutes a 'morally acceptable' solution is another issue which is side-stepped.

THE SCIENCE PARADIGM

The science paradigm embodies the scientific method which originates with Descartes and Rationalism and which has led both to the development of the 'hard' sciences such as physics and chemistry and to the role of science and technology in modern life.

The science paradigm has been described as a 'learning system' characterized by *reductionism, repeatability* and *refutation*.[2]

> 'We . . . *reduce* the complexity of the variety of the real world in experiments whose results are validated by their *repeatability*, and we . . . build knowledge by the *refutation* of hypotheses.'[2]

The steps of the scientific method have been described in many forms.

- *Observations* are made after the phenomenon to be studied is defined.
- A *hypothesis* is postulated within a theory which seeks to explain the relationship among the observed variables.
- An *experiment* is designed to test the hypotheses and 'validate' the theory.
- If the hypothesis holds, the theory is accepted as an *explanation* of the observed events.
- Generalizations, or *laws*, lead to *predictions* of future states.

The scientific method and its paradigm have been successful when applied to fields of 'organized simplicity' or of 'unorganized complexity' as found in domains of the physical sciences, the so-called 'restricted sciences'.

Two systems methodologies which embody the science paradigm are Operations or Operational Research, and Systems Engineering and the Classic Systems Research approaches.

Operations or Operational Research

Description of Field
Operations or Operational Research (OR) can be defined as the application of quantitive methods to decision-making, in particular, *the use of mathematical models for optimization.* Many procedures dealt with in management science, decision science and so on can also be called Operations Research. The field dates back to World War II when efforts were made to find optimal solutions for allocation, logistic and transportation problems in military operations.

Nature of Domain
In a general sense, OR applies to any domain. However, due to the very nature of its paradigm which defines precise system boundaries, the solutions obtained are only valid for the closed system over which the solution is formulated.

Problems to Which Suited
The method of OR is best suited for well defined and well structured problems such as are found at the *operational* or *tactical* level. This means that it is probably better suited to 'hard' rather than 'soft' systems, where the 'hard–soft' dichotomy refers to its physical–mechanistic versus its biological-behavioral content.

Application of the Paradigm
In OR, the science paradigm is slightly modified. Several alternatives (instead of a single *hypothesis*) are postulated as possible solutions or paths leading to achievement of the defined objectives. A '*decision model*' evaluates alternatives. The model is tested to check whether it is a good *explanation* of reality. The solution is evaluated by comparing system performance before and after the chosen alternative is incorporated in the system. The solution is then implemented and maintained.

Nature of Solution
Within the system defined by the assumed boundaries (or constraints) the model yields an optimum solution.

Criteria for 'Truth' Evaluation
In OR models, it is usual to formulate *objective functions* in terms of the *measures of effectiveness* embodying the modeller's goals, these being normally the goals of the decision-maker who is being helped by the study.

Proof of 'Truth'
The OR model is 'self-serving' or tautological, in that it provides the function by which the solution can be tested for optimality. It does not offer a guarantor *outside* the model, apart from the attempts at evaluating the correspondence between model and reality and the improvement between the old and new systems. (The lack of guarantor leads directly to the need for introducing a metasystem in which

the desirability and effectiveness of the proposed solutions can be arbitrated at a systems level above that of the model under study, i.e. at the *metasystem level*. We will return to this point later.)

Systems Engineering and Classic Systems Research Methodologies

Description of Field

The differentiation between this methodology and Operations Research is one of scope and of degree. If there is a difference, it resides mostly in the claims of its proponents or adepts. The methods of Operations Research are similar to those used in Systems Engineering and related systems methodologies to which the name Classic Systems Research methodologies has been given.[3] Systems Engineering (SE) looks at reality broadly and thus accepts the ill-structure of domains. In its pure form, the application of OR models can only be made to sell structured systems whose attributes are specific and well defined. SE claims to embrace 'the total system research problem'.[4] Thus SE had been defined as:

> An approach involving analysis planning and design with a view to selecting the alternative, policy or course of action by which the systems performance requirements can be met.[5]

Nature of Domain

In SE, the problem setting is enlarged to embrace ill-structured domains. The influence of general systems theory is beginning to be felt when it is claimed that 'one strives to look at the entire problem as a whole'.[5] In preliminary research stages, SE questions the 'total program of work' around a problem in order to assess 'the immediate area of need' as well as a 'range of environmental factors'.[6]

Problems to Which Suited

SE is suited for tactical problems as well as those with strategic implications. SE attempts to discern the existence of the 'soft' system components in a problem, but its degree of success in dealing with them must be rated poor.

Application of the Paradigm

The SE methodology can be characterized as the application of the Operations Research methodology at the level of the entire system and at that of each of the sub-systems. The sub-systems can be sub-problems of the main problem or the several phases of it. The iterative process by which the problem is formulated, objectives are defined, measures of effectiveness chosen, alternatives generated and evaluated prior to policy selection, still relies heavily on the traditional method.

Nature of Solution

Alternatives are ranked by using the criteria chosen for evaluation.

Criteria for Truth Evaluation
In older versions of SE, alternatives were ranked according to criteria involving costs, benefits and effectiveness. Recently more sophisticated multicriteria decision models incorporating worth functions have improved the previously narrow evaluative process by which alternatives were compared.

Proof of 'Truth'
Cost benefit analysis is open-ended and thus does not provide any guarantee of truth. Thus, the wide ambitions of Systems Engineering and of the Classic Systems Research methodologies have not been realised. Providing decision-makers with several alternatives overlooks the *criterion problem* (which criteria are to be chosen), *the problem of interpersonal comparison* (how do we compare individual preferences), the *problem of consensus* (how can preferences be aggregated), and the problem of *implementation* (will the chosen solution be accepted by recipients).

THE SYSTEMS PARADIGM

The systems paradigm is no less scientific in a broad sense than the science paradigm or the metasystem paradigm. All are scientific in that they employ rational models of reasoning: they all proceed either by induction or deduction from premises to conclusions. All emerge in the same tradition of rational western thought.

The systems paradigm takes into account the indivisibility of systems domains where 'organized complexity' prevails. It originates from concerns that the science paradigm, which was designed to deal with the physical world, breaks down when faced with living systems. These entities are characterized by openness, low separability and high interdependence. This precludes the possibility that parts be treated in isolation since the parts when aggregated, display *emergent properties.*[7]

Thus reductionism and the logico-analytical framework of the science paradigm must be modified and complemented by a new approach whose foundations can be found in the new discipline of general systems theory.

At present, there is no unique systems paradigm which can be offered as a counterpart to the science paradigm. Several authors have described and/or used approaches which constitute a step towards a systems paradigm. The systems paradigm offered by van Gigch can be criticized for using an approach which is rather closely bound to the traditional scientific method and science paradigm.[8] It also uses 'hard' systems engineering terms such as 'design', 'optimization', 'implementation', terms which are not representative of the 'new learning system' pioneered by Checkland.[9,10]

Whereas van Gigch considers the systems paradigm as a 'fluid cybernetic process' where systems design functions are applied to a problem, Checkland limits himself to finding basic (systems) inadequacies in a situation in which a problem is perceived as a prelude to defining desirable and feasible changes which could lead to possible improvements.

Applied Systems Analysis and Applied General Systems Theory

Description of Field

Applied systems analysis is the name I shall give to the methodology which has evolved in the work carried out by P. B. Checkland and his colleagues at the University of Lancaster, UK. *Applied general systems theory* is the name given by van Gigch[11] to his brand of the Systems Approach. The two methodologies have moved in the same direction, and the work done at the Open University, UK, by the Systems Course Team can also be associated with this school of thought which straddles the 'hard' systems methodologies represented by OR and SE described above and the soft systems methodologies which derive their orientation from the life sciences and which will be described in subsequent sections.

Nature of Domain

Checkland and van Gigch are attempting to cope with the unstructured or ill-structured problems of the real world which exhibit the 'emergent systems properties' of wholes. They assume that socio-technical systems are endowed more with the characteristics of biological than of physical systems. We are here dealing with systems of 'organized complexity' which cannot be dealt with by the traditional scientific method which assumes systems with high separability and low interaction among parts. Checkland tends towards a view of human behaviour and organizations as not necessarily goalseeking but rather as maintaining 'on-going relationships' in the context or *climate* of what Vickers[12] has called 'appreciative systems'. Thus he favours the description of a problem situation as having elements of *structure* which refers to the relatively slow-to-change features and the constantly-changing elements of *process*. The *climate-structure-process* approach constitutes the basis for a learning system by which the problem situation is interactively explored and perceived.[13]

Problems to Which Suited

This methodology, in particular Checkland's approach, is decidedly designed to tackle practical real world issues. The University of Lancaster group have applied their methodology to more than 100 systems studies. The Open University System Course Team have applied it to the study of systems performance and failures.[14]

Application of Paradigm

Checkland's version of the systems paradigm consists of a process by which conceptual models of systems relevant to the problem situation studied are built via *root definitions* which are descriptions of the six basic characteristics of the system, i.e. (1) ownership; (2) actor(s); (3) transformation process; (4) customer; (5) environment and (6) world view.

The conceptual models embody the *minimum essential activities* by which the systems can be represented. By comparing the models with reality and challenging the extent to which they are a faithful image of the present reality, modifications and changes to the system are suggested, designed and implemented.[15]

Van Gigch's version of the systems paradigm is described as a cycle or 'cybernetic fluid process' consisting of basic design functions by which soft systems can be fashioned. While claiming to owe its foundations to general systems theory, van Gigch's methodology suffers perhaps from incorporating too many vestiges of 'hard' systems engineering such as 'design', 'optimization', 'implementation' and the like. In contradistinction with Checkland[16,17] and Popper[18] who defend incrementalism and piece-meal engineering, van Gigch argues in favour of Popper's *Utopian Engineering*, that is, solutions at the level of the global system. Whether these solutions can be found or even formulated is not always clear. Checkland's more limited vision of the Systems Approach may have the decided advantage of practicality.

Nature of the Solution
Checkland's methodology yields partial desirable changes when the conceptualization stemming from the root definitions conflicts with the perceived present reality. In van Gigch's methodology a solution is the output of the systems design process applied at the scale of the total system.

Criteria for 'Truth' Evaluation
In Checkland's scheme, changes to the system must meet two criteria. They must be *desirable* and *feasible*. The more comprehensive approach advocated by van Gigch somewhat clouds the issue of the specific criteria to be met. He favours the use of multidimensional decision models where more than two criteria can be satisfied. This calls for compromise and trade-offs, a not too unrealistic representation of the real world. He also calls for the incorporation of ethical considerations in the calculus of system optimization. In Churchman's[19] terms, systems science must be 'self-reflective' and evaluate the results of its own deeds.

Proof of 'Truth'
For Checkland the choice of a desirable change stems from a comparison between the conceptual models stemming from root definitions and the world view of perceived reality. Thus *consensus* provides the guarantor of truth. Consensus is also invoked by van Gigch as a necessary ingredient of truth but its practical validity may be lost when dealing at the scale of the global system.

In summary, the methodologies dealt with in this section move systems science in the direction of holistic thinking and they are useful complements to the more traditional approaches described earlier. Checkland's approach goes a long way in answering the need for a practical methodology tailored to soft-systems domains. Its lack of theoretical foundation may prove a disadvantage in the future if more progress is to be made in conquering the methodological dilemmas confronting it.

The Methodology of Diagnosis

In this section two theories which exemplify the methodology of *diagnosis* are given. This methodology has been developed for the most part in the medical sciences and of late has been applied to other fields. For surveys refer to Pipino[20] and Bouwman.[21] It is clear that the methodology of diagnosis can be used in both contexts, the science and systems paradigms. It is presented at this point because the two theories chosen for illustration are holistic in nature and are in that sense representative of the systems paradigm.

The methodology of diagnosis proceeds as follows:

- Given an entity perceived as system, acceptable and unacceptable system *states* are defined.
- Features, attributes or patterns of behaviour are extracted to generate categories of classes.
- Observations from the system under study are assigned to classes defined earlier.
- In the *detection phase*, a decision is made whether the system is in an acceptable or unacceptable system state.
- In the *diagnosis and evaluation phase*, the unacceptable system state(s) is identified.
- In the *treatment phase*, the unacceptable system state is restored to one which is acceptable.
- In the *control phase*, a prediction or prognosis is attempted, to determine whether the system will remain in an acceptable state or will lapse back to an unacceptable one.[22]

The intent is to define a good *descriptive* model, that is, one which is a faithful representation of reality. The model is then used in a *normative* sense to describe how the system *ought to be* and against which an actual system or situation can be compared to detect unacceptable departures.

The crux of the successful application of this methodology resides in having a model which can be used in the *dual descriptive-normative role*. Two models which may potentially be used to achieve this purpose are described below:

(i) Miller's general systems theory of living systems, and
(ii) Autopoiesis.

A General Systems Theory of Living Systems and Autopoiesis

Description of the Field

1. *Miller's general systems theory of living systems.* J. G. Miller[23,24,25] has developed a general systems theory which describes isomorphisms across seven levels of living systems. These are cells, organs, organisms, groups, organizations, societies and supranational systems. He has conceptualized nineteen (19) basic energy-matter-information processing subsystems which he claims are common to all levels. This theory can be considered a problem-solving methodology to the extent that it provides a descriptive-normative model against which actual systems can be compared.[26]

2. *Autopoiesis.* Autopoiesis can be considered a more advanced general systems model of living systems which originates in theories of microbiology.[27,28] In the

same sense as Miller's theory, it can be termed 'a paradigm of inquiry' in that living systems can be studied, their properties simulated and the effects of specific environmental perturbations monitored by means of the model of the process of autopoiesis. Some proponents of autopoietic systems contrast Miller's model to theirs and claim his to be obsolete.[29,30]

Nature of Domains
The use of both Miller's theory and autopoiesis is concerned with the description of living systems whose processes can be studied in order to find identities and/or departures from those postulated in the models. Thus these models lie in the realm of 'soft systems' and may help us understand better the organizational and behavioural characteristics of the real world.

Application of the Paradigm
Not Discussed.

Problems to Which Suited
(i) Miller has developed his theory and model to encompass the seven levels of living systems. Thus, one could take systems displaying any or all of these levels and determine whether they function in the form presented by the model. The model probably offers most promise at the highest levels of complexity such as those exemplified by organizations, societies and supranational systems. As yet, we do not have very good descriptive models of these entities. Miller's characterization may be well suited to evaluate how socio-technical systems and their component sub-systems operate.

(ii) Autopoiesis is too new as a field of endeavour to judge how it will be developed and applied. Reference has been made to simulation studies by which the effect of environmental perturbations upon intricate phenomena occurring in the processes of cells can be studied and modelled.[29,30] Whether, as claimed, these processes can be transposed to higher levels to explain processes in higher living organisms is subject to conjecture.

Nature of Solution
In the context of the methodology of diagnosis, a solution (or 'truth') consists in recognizing the unacceptable state of a system and finding ways to restore it to an acceptable one. Thus a *solution* is obtained when the right diagnosis of ills and the correct treatment have been formulated. Good descriptive models to which actual systems in disarray can be compared, can offer potential for improving diagnosis. It is open to question whether they can provide hints for appropriate treatment.

Criteria to Evaluate 'Truth'
In diagnostic models, a 'solution' can be evaluated by:

- the costs involved in obtaining proper diagnosis;
- the extent to which unacceptable states depart from desired acceptable ones;
- the cost involved in restoring the system back to acceptable states.[31,32]

Proof and Guarantee of 'Truth'
A solution is reached when the system which starts from an unacceptable state is restored to an acceptable state. Thus the diagnostician is dependent on a good model which clearly discriminates among system states and whose patterns and features can be easily recognised. Of course, treatments may have unexpected effects *outside* the system. Who is the guarantor then remains an open question.

THE METASYSTEM PARADIGM

The metasystem paradigm originates in the premise that one cannot arbitrate deficiencies among systems in other than a metalanguage, that is in the language of a metasystem which lies at a system level *above* that of the systems whose control is sought. Beer exemplified the concept of a metasystem by suggesting isomorphisms between the physiology of the brain and the hierarchy of control used in the firm.[33] The concept of metasystem can also be found in the conceptual model of problem-solving systems ideals predicated by Klir,[34] Klir & Uyttenhove,[35] Ulrich[36] and earlier in Churchman.[37]

Description of the Field
A cybernetic metasystem model. In *Brain of the Firm*, Stafford Beer[38] suggested that the organization could be modelled as a hierarchy of control systems patterned after the neurophysiology of the brain. This was followed by *Platform for Change*[39] which can be considered a cybernetic metasystem model. It is *cybernetic* because it features feedbacks, and it is a *metasystem* because it represents a system to design other systems.

A conceptualization of problem-solving processes design ideals. The school of thought initiated by Churchman and exemplified by Mason,[40] Mitroff & Betz,[41], Mitroff & Sagasti,[42] Mitroff & Turoff,[43] Mitroff,[44] Mitroff, Barabba & Kilmann[45] and by Ulrich[46] integrates problem-solving methodologies into a hierarchy of problem-solving systems. This hierarchy is a general system model of design. It is devoid of content in the sense that it can be applied to all systems. It can be construed as the conceptualization of Beer's idea of the metasystem which can be used to design any other system.

Nature of Domain
The metasystem paradigm deals with epistemological questions. It is concerned among other things with the design of a paradigm by which all other paradigms are designed.

Problems to Which Suited
A design metasystem is devoid of content in that it addresses itself to the problem of judging other methodologies. It is not problem oriented but constitutes an inquiring system by which methodologies for solving problems are designed.

Application of the Paradigm

Klir[47] conceptualized a hierarchy of epistemological levels of systems which are differentiated by the level of knowledge about a set of variables in the system, about potential states associated with each variable and their constraints. 'A higher level system entails all knowledge of the corresponding systems at any lower level and contains some additional knowledge which is not available at lower levels.' Klir defines six levels.

- *The source system* where systems at this level are defined as sources of available data.
- *The data system*, when a source system is supplemented by data.
- *The generative system*, which provides a basis for generating data.
- *The structure system*, which consists of a set of generative systems or lower level systems embedded in the larger system and which display relations of coupling among its elements or components.
- *The metasystem*, a level at which systems are defined by a set of structure systems or lower level systems and an invariant procedure which describes transitions between those systems.
- *The meta-metasystem* containing all the metasystems and which can be defined by 'assuming an invariant metaprocedure describing transitions between the procedures of the metasystems'.

Klir and his co-workers have performed data systems experiments in which this conceptual framework has been operationalized in what is presently called Reconstructability Analysis.[48,49,50] By its very nature it is a form of systems modelling which must be considered metamodelling.[51,52]

Ulrich[53] conceptualized problem-solving as a process of social systems design, a process that is implicitly or explicitly oriented toward some presupposed design ideal. The ultimate design ideal is that of a *problem-solving system*, by which he understands 'a system which was given by its designer the capability to perform a range of *internal* problem-solving processes without any further intervention on the part of the designer.[53] The underlying idea is that social systems are not adequately considered from a purely functionalist point of view but must be understood (and designed for) as at least partly internally motivated systems, i.e. as *purposeful systems*. A purposeful problem-solving system must invariably perform three basic kinds of complementary problem-solving processes.[53,54]

(i) *Inquiry:* it must produce meaningful knowledge with respect to its purpose (it ought to be enabled to become a purposeful inquiring system).
(ii) *Action:* it must secure the purposeful use of this knowledge (it ought to become a purposeful action system).
(iii) *Valuation:* it must responsibly evaluate its production and use of knowledge from the point of view both of those to be served and those who are not served but may be negatively affected (it ought to become a purposeful-morally responsible valuation system).

By applying his basic distinction between internally and externally motivated systems to the three levels, he arrives at a taxonomy of six conceptual system designs and corresponding design ideals: responsive versus reflective inquiring systems, purposive versus purposeful action systems, and self-sustaining or 'housekeeping' versus morally responsible valuation systems.

His approach differs significantly from that of the before-mentioned authors in one respect; his intent is mainly, though not exclusively, a critical one against what he considers to be the technocratic (supposedly value-free) employment of the systems concept in 'systems science', managerial cybernetics, MIS design and similar approaches. For Ulrich, it is within the context of a *critical* systems approach – a systems approach that provides the problem solver with the heuristic support he needs to reflect on the value-laden presuppositions and consequences of a systems design – that reflection on the design ideals built into our designs takes on an important function. Since his original 1977 paper, he has considerably expanded the scope of his approach. He has published a comprehensive study on the epistemological and heuristic foundations of a critical systems approach, entitled *Critical Heuristics of Social Planning*[54]. Chapter 17 included in the present volume, represents the direction of his more recent work 'toward an openly but critically normative metasystem paradigm'.

In Beer's terms the organization as a total system is a hierarchical structure of systems which requires:

- **organization** to handle complexity;
- **science** to manipulate knowledge;
- **cybernetics** to redesign itself according to a continually changing model;
- **a metasystem measure** by which the outputs of subsystems can be evaluated;
- **tools** to handle the information flows by which organizations are managed;
- **ethics** to manipulate change with compassion.[55]

The concept of metasystem also appears in policy-making and planning science. Dror[56,57] explicitly distinguishes between metapolicy-making, policy-making and postpolicy-making. Metapolicy-making is described as the way to improve the policy-making system, that is, policy on how to make policy. Faludi[58] discerns three levels in planning theory namely metaplanning, procedural and substantive theories of planning. Some initial hypotheses about the theory of metaplanning are developed by Emshoff.[59] Designing the metasystem for organizational decision-making is presented in Kickert and van Gigch[60] (see Chapter 3).

Nature of the Solution Obtained

The metasystem paradigm incorporates epistemological characteristics which are lacking in the two paradigms discussed above. In particular it includes:

- a hierarchy of problem-solving levels in which higher systems levels can judge and rate solutions for lower systems levels;
- a framework to provide evaluation criteria in metalanguage terms, i.e. in a language appropriate to judge lower system solutions;
- a guarantor of truth at each system level, always expecting the last (or highest).

Criteria to Evaluate 'Truth'

Beer's work shows that we need metasystem criteria to judge lower system propositions. Decisions in one system must be reformulated in the metalanguage. The metasystem, for example, must evaluate how lower level systems apply technology to produce waste as well as wealth. It must also be in a position to evaluate the ethical implications of the various systems designs, an issue which is embodied in the Morality of systems (Churchman,[61] van Gigch[62]).

Proof of 'Truth'

Truth can only be evaluated in metalanguage terms and can only be judged from the vantage point of the metasystem. Neither of the two paradigms previously considered could guarantee the nature or validity of their solutions because they lacked a metasystem from which this could be done. The science paradigm can only judge optimality in the context of the very model which produced the solution. The systems paradigm relied on consensus, a mild form of review by a higher system. In the diagnostic system, evaluation was tautological in that departure from the norm was a function of the adequacy of the descriptive model which was turned into a normative one for purposes of comparison and evaluation – thereby an empty proof.

CONCLUSIONS

This study compares three paradigms by which we design methodologies for problem solving.

Attempts at designing a methodology suitable for soft systems domains overlook the need to deal with issues such as the choice of criteria by which solutions can be evaluated, and the need for a guarantor for truth.

In OR, solutions are only valid in the context of the model by which they are optimized. In Systems Engineering, the solutions are open ended. In present versions of the systems paradigm allusions to consensus and ethics show the tendency to add the requirement that higher level systems need to evaluate lower level ones. The introduction of the metasystem paradigm partially answers these criticisms. However, even within this context, it may not be possible to admit a higher level guarantor beyond the highest system conceptualized. This is a major paradox for systems thinking, as Churchman[63] and François (see Chapter 5) have pointed out.

REFERENCES

1. T. Kuhn, *The Structure of Scientific Revolutions* (Univ. Chicago Press, Chicago, Ill., 2nd Ed., Rev., 1970).
2. P. B. Checkland, 'Science and the Systems Paradigm', *Int. J. of General Systems* **3**, 127–134 (1976).
3. Open University, *Systems Performance: Human Factors and Systems Failures Technology*. A Third Level Course, TD 342-1, and Systems Summer School 1976, *Student Notes* (The Open University, Milton Keynes, 1976).

4. J. E. McGrath, P. G. Nordlie and W. S. Vaughan, 'A Systematic Framework for Comparison of Systems Design Methods'. In: Optner S. L. (Ed.), *Systems Analysis* (Penguin, Harmondsworth, England, 1973), pp. 76–79.
5. E. S. Quade and W. I. Boucher (Eds.), *Systems Analysis and Policy Planning – Applications to Defense* (Elsevier, New York, 1968), pp. 11–14, 33–34.
6. A. D. Hall, *A Methodology for Systems Engineering* (Van Nostrand, New York, 1962), pp. 7–11.
7. P. B. Checkland. *Op. Cit.*
8. J. P. van Gigch, *Applied General Systems Theory* (Harper and Row, London and New York, 2nd Ed., 1978).
9. P. B. Checkland, 'Towards a Systems-based Methodology for Real-World Problem Solving', *J. Systems Engineering* **3**,2 (1972). Reprinted in G. Peters and J. Beishon (Eds.), *Systems Behaviour* (Harper and Row, London and New York, 3rd ed., 1982), and P. Checkland, 'Science and the Systems Paradigm', *Op. Cit.*
10. P. B. Checkland, *Systems Thinking, Systems Practice* (Wiley, London and New York, 1981).
11. J. P. van Gigch, *Applied General Systems Theory, Op. Cit.*
12. G. Vickers, *The Art of Judgement* (Basic Books, New York, 1965).
13. P. B. Checkland, 'Science and the Systems Paradigm', *Op. Cit.*
14. Open University, *Op. Cit.*
15. D. S. Smyth and P. B. Cleveland, 'Using a Systems Approach: The Structure of Root Definitions', *J. Applied Systems Analysis* **5**, 1 (1976).
16. P. B. Checkland, 'Towards a Systems Based Methodology . . .', *Op. Cit.*
17. P. B. Checkland, 'Science and the Systems Paradigm', *Op. Cit.*
18. K. R. Popper, *Poverty of Historicism* (Routledge and Kegan Paul, London, 1957).
19. C. W. Churchman, *Challenge to Reason* (McGraw-Hill, New York, 1968).
20. L. L. Pipino, 'The Application of Fuzzy Sets to System Diagnosis and the Design of a Conceptual Diagnostic Procedure', *Doctoral Dissertation* (Univ. of Massachusetts, Amherst, Mass., 1975).
21. M. J. Bouwman, 'Human Diagnostic Reasoning by Computer: an Illustration from Financial Analysis', *Management Science* **29**, 3-672 (1983) and 'Expert Versus Naive Decision Making in Accounting: A Summary', *Accounting Organisations and Society* **9**, 325–327 (1984).
22. L. L. Pipino. *Op. Cit.*
23. J. G. Miller, 'The Nature of Living Systems', *Behavioural Science* **20**, 345–365 (1975).
24. J. G. Miller, 'The Nature of Living Systems', *Behavioural Science* **21**, 295–468 (1976).
25. J. G. Miller, *Living Systems* (McGraw-Hill, New York, 1978).
26. K. J. Vandervelde and J. L. Miller, 'The Urban Grant University Concept: A Systems Analysis', *Behavioural Science* **20**, 273–295 (1975).
27. F. G. Varela, H. R. Maturana and R. B. Uribe, 'Autopoiesis: The Organization of Living Systems, its Characterization and a Model', *Bio-systems* **5**, 187–196 (1974).
28. H. R. Maturana, 'Organisation of the Living: A Theory of the Living Organisation', *Int. J. Man-Machine Studies*, **7**, 313–332.
29. M. Zeleny, 'Self-Organization of Living Systems: A Formal Model of Autopoiesis', *Int. J. of General Systems* **4**, 13–28 (1977).
30. M. Zeleny (Ed.), *Autopoiesis: A Theory of Living Organization* (North-Holland, Amsterdam, 1981).
31. A. G. Gorry and G. O. Barnett, 'Experiences with a Model of Sequential Diagnosis', *Computers and Biomedical Research* **1**, 490–507 (1968).
32. A. G. Gorry and G. O. Barnett, 'Sequential Diagnosis by Computer', *J. Amer. Medic. Assoc.* **205**, 849–854 (1968).
33. S. Beer, *Brain of the Firm*, 2nd Ed. (Wiley, New York, 1981).
34. G. J. Klir, 'Identification of Generative Structures in Empirical Data', *Int. J. General Systems* **3**, 89–104 (1976).

35. G. J. Klir and H. J. J. Uyttenhove, 'Computerized Methodology for Structure Modelling', *Annals of Systems Research* **5**, 29–66 (1976).
36. W. Ulrich, 'The Design of Problem-Solving Systems', *Management Science* **23**, 1099–1108 (1977).
37. C. W. Churchman, *The Design of Inquiring Systems* (Basic Books, New York, 1971).
38. S. Beer, *Brain of the Firm, Op. Cit.*
39. S. Beer, *Platform for Change* (Wiley, New York and London, 1975).
40. R. O. Mason, 'A Dialectical Approach to Strategic Planning', *Management Science* **15**, B-403 – B-414 (1969).
41. I. I. Mitroff and F. Betz, 'Dialectic Decision Theory', *Management Science* **19**, 11–24 (1972).
42. I. I. Mitroff and F. Sagasti, 'Epistemology as a General Systems Theory: An Approach to the Design of Complex Decision-Making Experiments', *Philosophy of the Social Sciences* **3**, 117–134 (1973).
43. I. I. Mitroff and M. Turoff, 'Technological Forecasting and Assessment: The Whys Behind the Hows', *IEEE Spectrum* **10**, 62–71 (1973).
44. I. I. Mitroff, 'Towards a Theory of Systemic Problem Solving Prospects and Paradoxes', *Int. J. General Systems* **4**, 47–59 (1977).
45. I. I. Mitroff, V. P. Barabba and R. H. Kilmann, 'The Application of Behavioural and Philosophical Technologies to Strategic Planning: A Case Study of a Large Federal Agency', *Management Science* **24**, 44–58 (1977).
46. W. Ulrich, *Op. Cit.*
47. G. J. Klir, 'Identification of Generative . . .', *Op. Cit.*
48. G. J. Klir and H. J. J. Uyttenhove. *Op. Cit.*
49. G. Broekstra, 'Constraint Analysis and Structure Identification', *Annals of Systems Research* **5**, 67–90 (1976).
50. G. Broekstra, 'Constraint Analysis and Structure Identification II', *Annals of Systems Research* **6**, 1–20 (1977).
51. *Int. J. General Systems*. Special Issue on Reconstructibility Analysis **7**, 1 (1981).
52. R. E. Cavallo and G. J. Klir, 'Decision Making and Reconstructibility Analysis', *Int. J. General Systems* **8**, 243–255 (1982).
53. W. Ulrich, *Op. Cit.* and Personal Communication (May, 1982).
54. W. Ulrich, *Critical Heuristics of Social Planning: A Reconstruction of Kantian A-Priori Science for Planners and Practical Philosophers* (Paul Haupt, Berne, Switzerland, 1983). See Excerpt in this text, Chapter 17.
55. S. Beer, *Platform for Change* (Wiley, New York and London, 1975).
56. Y. Dror, *Public Policy Making Re-examined* (Chandler, Scranton, Penn., 1968).
57. Y. Dror, *Design for Policy Sciences* (Elsevier, New York, 1971).
58. A. Faludi, *Planning Theory* (Pergamon Press, Oxford, 1971).
59. J. R. Emshoff, 'Planning the Process of Improving the Planning Process: A Case Study in Meta-Planning', *Management Science* **24**, 1095–1108 (1978).
60. W. J. M. Kickert and J. P. van Gigch, 'A Metasystem Approach to Organizational Decision Making', *Management Science* **25**, 1217–1231 (1979). See Chapter 3, this text.
61. C. W. Churchman, *Challenge to Reason* (McGraw-Hill, New York, 1968).
62. J. P. van Gigch, *Applied General Systems Theory, Op. Cit.*
63. C. W. Churchman, *Challenge to Reason, Op. Cit.*

SECTION II

THE METASYSTEM PARADIGM

3. A METASYSTEM APPROACH TO ORGANISATIONAL DECISION-MAKING*

Walter J. M. Kickert and John P. van Gigch

The metasystem approach to organizational decision-making presented in this paper is set in a control systems framework. Decision-making is approached from a control system viewpoint. the concept of metasystem is introduced as a framework for decision-making about decision-making that is, the structuring of a decision-making process and the determination of organizational objectives. The metasystem framework is illustrated by means of examples drawn from a case study.

INTRODUCTION

The study presented in this paper forms part of a broader study of the structure and structuring of organisational decision-making processes.[1] In a previous paper[2] the importance of structural aspects of organisational decision-making in explaining these decision processes was stressed. It was shown that the usual concept of rationality is inadequate and that dynamic and other structural aspects should be incorporated in the concept of rationality in order to explain the structure and process of decision-making. These ideas were elaborated by introducing a new type of decision-making, namely 'structural' decision-making. Interest here is in the structure and structuring of the decision-making, that is, the process leading to the decisions about the organisation of the decision-making process.

This concern for structural decision-making also appears in the literature. The theory of decision-making in organisations centering around phase-models is concerned with the procedure of the decision-making process. Phase-models describe the pattern adopted by the sequence of systems during the decision-making process. However, these models restrict themselves to only one dimension of the structure of the process, namely time. It is clear that organisational decision-making processes have a much more complex structure and that the consideration should not be limited to a one-dimensional phase pattern. Other obvious dimensions are those which relate to the complexity of the relations between persons, groups and organisations. There are several kinds of theories which deal with this aspect of structure, such as the classical organisational theories known as 'Scientific Management' and 'Scientific Administration', the theories dealing with line-staff structures, functional structures, horizontal coordination, etc. The political theories about power and influence also fall in the category of theories dealing with relations between actors in decision-making. In reference 2, a descriptive framework was proposed for the structure of a decision-making process using a systems theoretical approach, which by its generality includes most of the properties and

* Reprinted with permission from: *Management Science*, **12**, 1217–1231, 1979. Copyright 1980, The Institute of Management Science.

variables of the above-mentioned theories. This structural framework describes an organisational decision-making process as a path through the three-dimensional space spanned by sub-, aspect and phase systems. The structure of this process consists of the relations between all parts of the system, that is, the transfers from one part of the system to another. In this paper a first step towards a theory about the organisation of decision-making is made by investigating the relationship of such a theory within its wider framework. Here a conceptual framework for decision-making is presented from a control system viewpoint, in which the concept of structural decision-making is embedded in a metasystemic model. This forms the basis for a theory about the organisation of decision-making itself, that is, a theory on how to explain and organise decision-making. The result of this broader study has been published.[3]

THE CONCEPT OF A METASYSTEM

'Meta' stems from Greek where its meaning is 'after'. Metaphysics start where physics end, that is, it comes after physics. 'Meta' is nowadays mostly used in the meaning of 'above'. Methodology is the science of scientific method. Methodology stands above science and is therefore considered as a metascience. In order to describe the metascientific properties of methodology one usually distinguishes between object-language and metalanguage. A sentence like '1 + 1 = 2' is an object-language sentence. A sentence like ' "1 + 1 = 2" is an arithmetical equation' but is, however, a metalanguage sentence. The sentence no longer deals with the objects but with a higher-level property. It stands above the first sentence and it is in this sense that the concept of a metasystem is used. A system is defined as a set of elements and relationships between the elements. As long as one deals with these elements and relationships – the objects of the system – the considerations on the system are at object level. When this level is exceeded, the considerations are at metasystemic level.

The literature of organisation science and systems theory refers in several places to the concept of a metasystem, a system placed at a level above the system under consideration.

Beer[4,5] explicitly refers to the metasystem when comparing the neuro-physiology of the brain to the organisation of the firm. Later, the idea of a metasystem is elaborated into a cybernetic metasystem model. The concept of a metasystem has implicitly or explicitly been incorporated in the design of computer hardware and software.

Churchman's[6] incursion into the design of inquiring systems is premised on the need for designing the metasystem for human inquiry. Churchman explicitly states the need not only for management science, but also for a science of management, a decision-making science from which we can learn to decide on how to decide, that is, to make metadecisions.

Klir[7,8] conceptualized a hierarchy of epistemological levels of systems which are differentiated by the level of knowledge regarding the set of variables and of potential states which is contained at each level. Mathematics is considered the metalanguage of science. According to Rapoport,[9] mathematics serve to bring out the isomorphisms existing across particular sciences and thus acts as the metalanguage of General Systems Theory. Within mathematics one can refer to more specific metalanguages. Thus probability theory has been called the metalanguage of uncertainty and fuzzy set theory the metalanguage of ambiguity.[10–12] Recently Van Gigch[13] discussed the need for a metasystem in the context of a comparison of methodologies for systems design and problem-solving. The concept of metasystem also appears in decision-making science, particularly the related fields of policy-making and planning science. Dror[14] explicitly distinguishes between metapolicy-making, policy-making and postpolicy-making. In a later work[15] Dror adds one more level, namely the level of megapolicy-making which involves the determination of the postures, assumptions, and main guidelines to be followed by specific policies. Metapolicy-making is described as the way to improve the policy-making system, that is, policy on how to make policy. Faludi[16] distinguishes between three levels in planning theory namely metaplanning procedural theories and substantive theories on planning. The first deals with the design of the planning system, the second with the way plans operate and the last with the area of concern. Faludi explicitly defines metaplanning as the design of planning agencies and their procedures. Some initial hypotheses about a theory of metaplanning are developed by Emshoff[17]. Mitroff and Betz[18] explicitly refer to metadecision-making.

As will be clear from this brief and surely incomplete review of the concept of metasystem in some of the literature, the concept is used in many other different ways. The interpretation of the concept adopted here is related to the first-mentioned methodological approach which stems from the difference between object-level language and metalevel language. Let us now proceed to discuss how it can be viewed in relation to organisational decision-making and control.

A CONTROL SYSTEM APPROACH TO DECISION-MAKING

In this section we will show how the adoption of a control systems point of view can lead to a better insight into decision-making particularly about the conceptual differences between various sorts of decision-making encountered in literature, such as structural, procedural, operational and substantive decision-making. One may find more or less vague, confusing and tautological definitions of these terms. Structural decision-making is concerned about the structure, that is, the organisation of the decision-making process. Operational decision-making deals with how decision-makers operate, procedural decision-making with the procedures used in decision-making – which is almost synonymous to the second term – and substantive decision-making deals with the area of concern. Although these concepts might well be very useful in certain cases like the field of planning (see[19]) we feel that a clearly defined conceptual framework in which all these various concepts fit in a consistent way is mostly missing and we think that a control systems viewpoint may serve to construct such a framework. Moreover, we will show that this viewpoint leads to a model in which not only the decisions are emphasized but in which also attention is paid to the important issue of the implementation of decision, that is, the changes that the decision brings about.

We will introduce and clarify the control systems approach by beginning with the well-known classical model of *homo economicus*.

In a classical model of *homo economicus* the decision-making is represented by:

- a set of actions A,
- a set of states (outcome, effects) S,
- a relation $R : A \rightarrow S$ indicating which action results in which next state, that is a model of the situation under consideration,
- a preference ordering over the possible states, represented by a value function V.

The decision-maker is supposed to choose the action which results in the most preferred outcome (state), that is, to take an optimal decision. There are several basic aspects of this model that can be noted. First, the decision is the choice out of a set of actions. Second, the interest of the decision-maker is with the possible effects of the alternative decisions (actions). The decision-maker wants to improve some situation, that is, he strives for a desired situation, given that he has a preference ordering (goal) over the possible situation values (the states). From this goal he will derive the best action which can be taken knowing what action will have which effect, for he has a model of the situation he tries to change, the known relation R. A comparison between this model of decision-making with the usual model of a control system clearly shows the parallels: the decision-maker is the controller, his control actions are the alternative decisions from set A which are fed into the system to be controlled, the controlled system. This latter system is modelled by the relation R between actions A and states S. The preference ordering (goal) results in a preferred state from set S which the decision-maker (controller) tries to attain by choosing an optimal action from set A. The main objective of decision-making is to change and improve some situation of a system by means of control. It is clear that in this view implementation can never be overlooked.

According to the 'control paradigm' of de Leeuw[20] any interesting phenomenon can be modelled by means of a control-system configuration. A control system is defined in a very general way. A control system consists of a controller (CR), a controlled system (CS) and an environment (E). Control is defined as any form of directed influence of the controller on the controlled system. The union of controller CR and controlled system CS is called the control system C (see Figure 1). Note that in this general framework the definition of control is broad and not solely restricted to the direct control of states or outputs of the controlled system. In this paradigm phenomena are considered from the point of view of directed change. The behaviour of the controlled system CS is influenced by the environment E and by the control actions of the controller CR. Control by CR of CS is intended to bring about a behaviour of CS which is desired by CR, that is, control is based on a goal. The definition of control as 'any form of directed influence' embraces the goal concept. Without a goal there is no directed influence.

Besides the multitude of viewpoints that can be obtained by different interpretations of a certain situation in terms of controller, controlled system and environment, de Leeuw[20] has also elaborated a general classification of possible modes of control by *CR*.

The controlled system *CS* is influenced by the environment *E*. Hence *E* indirectly influences the controller *CR* as well. In order to obtain a desired behaviour of *CS* the controller *CR* can therefore in principle also exercise influence on the environment *E*. In addition to the direct control of *CS* by the controller *CR* – called 'internal' control –, another mode of control is to influence *E* in order to obtain a desired behaviour of *CS* indirectly – called 'external' control.

Consider the controlled system *CS* as a black box, consisting of an environmental input x, a control action input u, and a transfer function f which determines the output y. The controlled system *CS* can then be described by the equation $y = f(x, u)$.

Assume that the goal of the system can be considered as a subset G of the set of all possible outputs Y. Based on this goal and the knowledge about the controlled system *CS*, the controller *CR* has to choose appropriate control actions. From this representation of *CS* it is clear that the controller can bring about a desired behaviour of *CS*, that is, the attainment of goal G, in the following ways:

(1) choose an appropriate control action u. This is the normal type of control. Nothing else is changed;
(2) change the structure f. It is clear from the black-box model that this will affect y and thus indirectly affect the goal attainment;
(3) change the goal G. Obviously a change in goal directly influences the goal attainment;
(4) change the input x. This is the external mode of control. Influencing E will indirectly influence y and hence the goal attainment.

In reference 20 the first three modes of control are called (1) 'routine' control, (2) 'adaptive' control and (3) 'strategic' control, respectively. Considering the fact

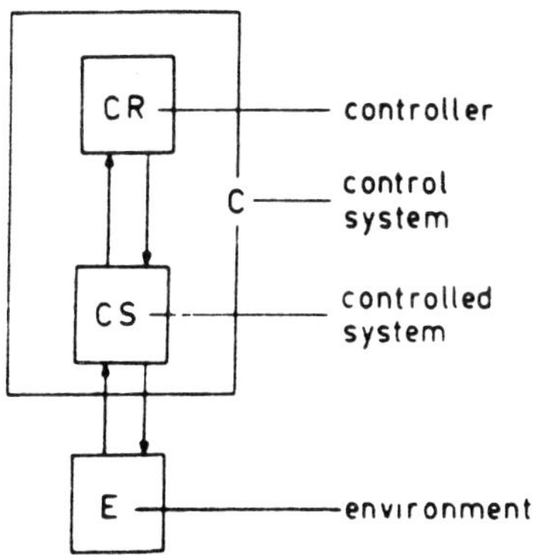

Fig. 1. General Model of Control System.

that these three modes of control can also be applied to the environment if one perceives it as a black box too, this results in a control characteristic consisting of six modes of control $\langle IR, IA, IG, ER, EA, EG \rangle$, where *I*- means internal, *E*-external, -*R* routine, -*A* adaptive and -*G* goal control. This characterizes a general typology of possible control modes.

Finally we want to introduce the concept of metacontrol in terms of this paradigm. Metacontrol is the control of the control, that is, the directed change of the controller itself in order to improve its control. The controller is controlled by a metacontroller, that is, at a next higher level, the controller has become the controlled system of the metacontroller. So there are two levels of control, the control at the object-level of *CS* by *CR*, and the control at the metalevel of *CR*, by the metacontroller meta *CR* (see Figure 2). Metacontrol can be dealt with by applying the control paradigm at a next higher level. The controller *CR* becomes a new controlled system CS' controlled by meta-$CR = CR'$. This means that the environment E' of the metacontrol system $CS' \cup CR'$ is generally not identical to the original environment *E* of the control system $CS \cup CR$. Note that in fact, one of the bases for differentiating control systems in organisation is a differentiation of the environment. Implicitly, this is the basis of differentiation, as drawn by managerial authors, or organisation control systems into 'strategic', 'tactical' and 'operational'. The same control characteristics can be applied to the metalevel, yielding six modes of control $\langle IMR, IMA, IMG, EMR, EMA, EMG \rangle$, where the additional –*M*– refers to the metalevel.

For an extensive treatment of this control-systems framework see the original publication of de Leeuw[20] and subsequent publications.[21,22]

We will now try to explain the concept of 'structural' decision-making in terms of this control-systems framework.

We have shown the analogy between decision-making and control. The decision-maker or the decision-making process could be identified with the concept

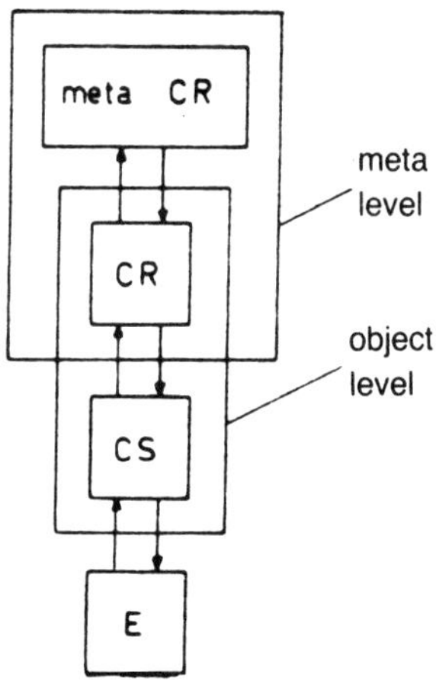

Fig. 2. Control and Metacontrol.

of controller. Therefore interest in the structure of the decision-making process means interest in the structure of the controller. Note that this is different and distinct from concern with the structure of the controlled system. Structuring the controlled system would be a particular kind of control by the controller, namely an 'internal adaptive' control mode. The decision-making aim might well be to change the structure of a particular situation. However, if the intention is to change the structure of the controller itself this can only be a structural mode of control performed by a controller at a higher level, that is, an 'internal adaptive' control performed by a metacontroller. In the case of structural decision-making we are concerned with the structure of the decision-making process which itself is the outcome of a decision-making process on a higher level. By analogy with the above-mentioned terminology we call decision-making which is concerned with the objects of decision-making, namely the decisions themselves, decision-making at the object-level, whereas decision-making concerning the structure of the decision-making process is called decision-making at the metalevel.

A METASYSTEM APPROACH TO ORGANISATIONAL DECISION-MAKING

Consider decision-making in the control-systems framework. The situation about which the decisions are made and where they must be implemented is viewed as the controlled system *CS*. The decision-maker or the decision-making process is viewed as the controller *CR*. As stated above, the union of controller and controlled system constitutes the control system *C*. Furthermore, influences from the environment *E* on the situation must be considered. This control-system viewpoint is made relative to the level of consideration, that is, the same framework can be used at different levels. We will use the index i for the object-level and the index $i + 1$ for the metalevel.

Now let us consider the three alternative forms of metacontrol which can directly be deduced from the *E*, *CS*, *CR*, meta-*CR* configuration. The metacontroller meta-*CR* does not have to restrict its metacontrol to *CR*. Via external metacontrol the metacontroller can influence *CS*, as well as the combination of *CS* and *CR*. (We will not consider metacontrol of *E* or *E'* here.) Let us consider these three possibilities somewhat more closely. In the first case the metacontroller CR_{i+1} has as its metacontrolled system CS_{i+1} the object-level controller CR_i, that is CR_i is equivalent to $CS_{i+1}(CR_i \Leftrightarrow CS_{i+1})$ as depicted in Figure 3. Note that this is identical to the configuration of Figure 2 which illustrated the structuring of the decision-making process.

The second alternative configuration is the case where the metacontroller CR_{i+1} exercises external control of the original controlled system CS_i, that is, where CS_i is equivalent to $CS_{i+1}(CS_i \Leftrightarrow CS_{i+1})$ as depicted in Figure 4. Imagine, for example, the case where a superior goes over his subordinate's head to directly influence the latter's subordinate. In this case the metacontroller CR_{i+1} and the controller CR_i together control the controlled system CS_i. Observe, however, that some doubt can arise as to the question whether this is still metacontrol. CR_{i+1} and

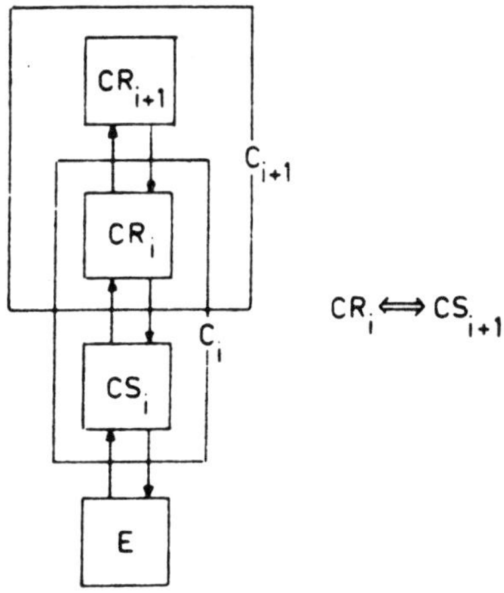

Fig. 3. One Possible Metacontrol Configuration.

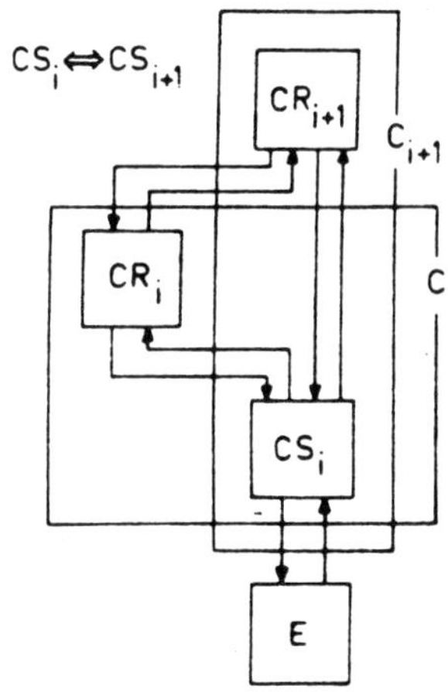

Fig. 4. Another Possible Metacontrol Configuration.

CR_i are simultaneously controlling CS_i and there does not seem to be any difference in level left. Both controllers now perform their tasks in a parallel way on the same level. So it completely depends on the question whether the CR_{i+1} control actions can be defined positively as actions of a higher metalevel in comparison with the CR_i object-level action, whether CR_{i+1} can be regarded as a metacontroller in this case.

The third alternative metacontrol configuration is where the metacontroller CR_{i+1} exerts influence on the combination of object controller CR_i and controlled system CS_i, that is CR_{i+1} controls the control system C_i as a whole. In this case CS_{i+1} is equivalent to $C_i(C_i \Leftrightarrow CS_{i+1})$ as depicted in Figure 5. The most obvious specification of this metacontrol of the combination of CR_i and CS_i in addition to

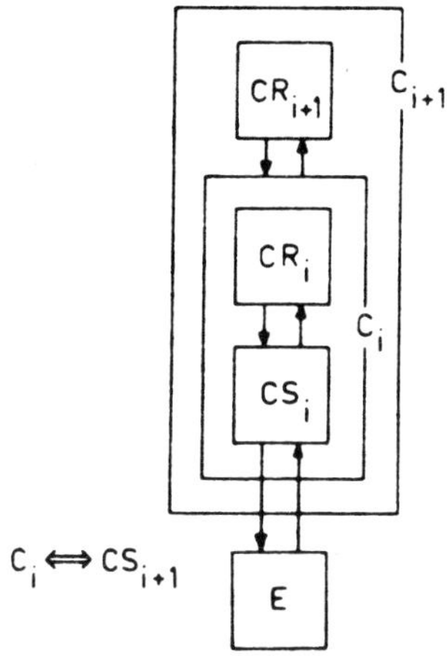

Fig. 5. A Third Possible Metacontrol Configuration.

the separate metacontol of CR_i and CS_i is the metacontrol of the relationships between CR_i and CS_i.

In brief, the three possible metacontrol configurations can be characterized as follows. The metacontroller CR_{i+1} controls the metacontrolled system CS_{i+1}. Now CS_{i+1} can be either equivalent to CR_i or to CS_i or to the union $CR_i \cup CS_i$. This enumeration is complete, apart from the possible environmental configurations.

There is probably no need to give an example of the first configuration ($CR_i \Leftrightarrow CS_{i+1}$) for it is in fact the concept of 'structural' decision-making which started the whole metacontrol treatment off. Influencing the structure of the decision-making process is an example of metacontrol of the object-level controller. An example of the second metacontrol configuration ($CS_i \Leftrightarrow CS_{i+1}$) is the case before-mentioned, in which a subordinate is not only controlled by his direct superior but also by his superior's superior. This metacontrol can be either supplementary to control by his direct chief or a replacement of that control. An example of the last-named metacontrol configuration ($C_i \Leftrightarrow CS_{i+1}$) is the case where it is not the structure of the decision-making process which is of primary concern, but the process of implementing the decisions, particularly the relationships between the decision-making and the implementation. Given that decision-making has been identified as the controller and implementation occurs in the controlled system, the relationship between decision-making and implementation is an example of the relationship between CR_i and CS_i.

The introduction of the metacontroller concept permits decision-making to be studied at three systemic levels:

(i) the lower level, that of the controlled system,
(ii) the middle level, that of the controller,
(iii) the higher level, that of the metacontroller.

Note that it is this middle level that we have called decision-making at the object level earlier. Studying decision-making at the 'lowest' level means studying the controlled system, to try to find out how the system behaves, in other words, to find a model of the system. In broader terms, it amounts to finding a description of reality. It is this concept that comes closest to the before-mentioned concept of 'substantive' decision-making which should 'serve to understand the area of concern'.[23]

There are indeed, some parallels between the before-mentioned distinctions made by Dror[24] and Faludi[25] and this three-partition of levels. Faludi distinguishes between metaplanning (defined as the design of planning agencies and their procedures), procedural theories (serving to understand the way planners operate) and substantive theories (serving to understand the area of concern). As stated above, the last type seems to coincide with our concept of lower level decision-making. Dror[26] distinguishes between metapolicy-making, policy-making and post-policymaking. The differences between these two partitions and ours mainly lies in the generality and coherence of our concepts. For instance, 'meta' with Faludi is restricted to structural design and post-policymaking, with Dror, is restricted to implementation and execution. Moreover the basis from which they arrive at their concepts and thus how these last are interrelated is not clear. In our conceptual framework metadecision-making is namely more general than structural decision-making only. Structural decision-making is merely one mode of metadecision-making, along with many other possible modes of metadecision-making. Following the before-mentioned general classification of control into the control of goals, the control of structure and 'routine' control, it will be clear that the same three-way partition in control modes applies to the metalevel. A clear example of a form of metalevel decision-making with the main emphasis on the goal control mode is the Dutch version of PPBS, that is the COBA system. In this system a hierarchically ordered system of goals is constructed in order to improve policy-making. The well-known means-end hierarchy is also an example of metalevel goal decision-making.

So far we have adopted the three-way partition in control modes to derive a typology of modes of metadecision-making. Obviously, depending on the kind of descriptive framework one adopts, different types of metadecision-making will arise. An example of another type of metadecision-making is the rationality of the decision-making process. To choose in favour of a rational-comprehensive or a disjointed-incrementalist mode of decision-making[27] or some intermediate mode such as mixed-scanning[28] is clearly a metadecision. Other modes of metadecision follow from other typologies such as, the three dimensional classification of Faludi[29] into (i) the blueprint versus process mode of planning, (ii) the rational-comprehensive versus disjointed incrementalist mode of planning, and (iii) the normative versus functional mode of planning or the six decision-making models of Nutt.[30]

The latter stipulates that the choice among these models can be made on the basis of an assessment of the contextual variables which affect the problem. However, there are no criteria external to the models presented by which an impartial judgement can be made. A decision made at the level of the universe of discourse of the six models themselves without recourse to metasystem criteria will of necessity be considered arbitrary and without rational basis. The criteria for choice and the methodology by which the choice is made can only be ironed out at another level where the differential weighting of the criteria can be the subject of a methodology expressly designed for that purpose. Therefore, what is needed is to design a decision-making procedure by which the methodology of choosing among competing models can be resolved.[31] This was also suggested by Heiskanen[32] who explicitly spelled out the need for a metasystemic methodology to evaluate different approaches and scientific strategies of theory formation in the social sciences.

STRUCTURAL METADECISION-MAKING

As stated in the introduction our main reason for developing the metaframework was our concern for structural decision-making. So let us now have a closer look at this mode of metadecision-making. As was explained earlier, according to de Leeuw[33] the three-way partition in different modes of control follows from the black box concept. In a black box, structure is defined as the input–output function. Here a slightly different definition of the concept of structure, namely the set of relationships S between the elements X of a system $\langle X, S\rangle$, will be adopted in order to visualise structural decision-making. Consider the complex multilevel decision-making process from Figure 6 (the environment has been omitted for the sake of convenience).

If the decision-making at the ith level CR_i controls the relationships between the elements A_i and B_i of the controlled system CS_i, this is called a structural mode of decision-making (control). Notice that this is still control at the object level. If the metadecision-making CR_{i+1} controls the relationship between the decision-making systems C_i and C_i' both objects of the meta-controlled system CS_{i+1}, this is a structural mode of metadecision-making from the point of view of CS_i and CS_i'. From the point of view of C_i and C_i' this structural control reduces to object-level

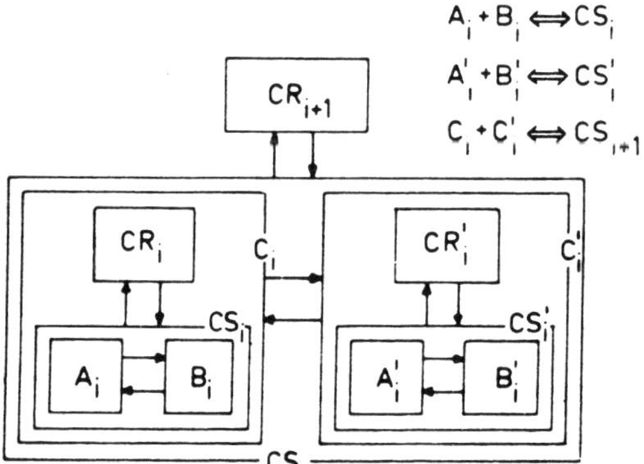

Fig. 6. Complex Multilevel Decision Making.

control. It should therefore be emphasized that meta- or object-level control completely depends on the level of consideration one adopts. What is metacontrol at one level reduces to object-level control at another and vice versa. We consider this inherent pluralism, which is due to the fact that the conceptual framework is object-independent, not as a negative ambiguity but as a fruitful property. Approaching the same problem from various points of view increases the possibility of fruitful problem-solving, an opinion which in methodology is strongly promoted by Feyerabend.[34]

If one drops the black box definition of structure as the input–output transfer function but adopts the definition as the set of relationships between the elements of the system, one might introduce further subclassification of modes of structural decision-making by classifying the concept of structure. In reference 35, we namely proposed to divide the structure of a decision-making process into three dimensions: sub-, aspect and phase systems. A system is defined as a set of objects W, a set of relations R_W between all these objects, an environment $E(W)$ with relations $R_{E(W)W}$ between E and W. Now a subsystem is defined as a subset of the objects W with all the original relations R_W, an aspect system is defined as all the objects W with only a subset of the relations R_W and a phase system is defined as a system identical to the original system only during a certain time interval.[36]

If we consider the participants of an organisational decision-making process as the objects of the system, then the partitioning of that system into sub-, aspect and phase systems can be interpreted as follows:

subsystems – groups, departments, the WHO
aspect systems – the issues, topics, the WHAT
phase systems – the phase, the WHEN.

The structure of this system is then defined as the set of relationships between sub-, aspect and phase systems:

relations sub/sub – interactions, power, communication
relations aspect/aspect – functional co-ordination issues, etc.
relations phase/phase – what sequence?
relations aspect/sub – who does what?
relations phase/sub – who acts when?
relations phase/aspect – what is dealt with when?

In other words, the structure, that is, the relations between sub-, aspect and phase systems should be interpreted as to 'who is doing what and when'. With this descriptive framework an organisational decision-making process is modelled as a path in the three dimensional space spanned by sub-, aspect and phase systems. The structure of this process consists of the relations between all possible blocks of the system.

Observe that we did not distinguish between prescriptive and descriptive theories of decision-making. This implies that the organisation of decision-making can mean two things: either the design of the structure or the description and explanation of the structure of a decision-making process. Although this is a very important distinction we will not dwell upon this theme here. For an extensive treatment of the distinction see Reference 37. Suffice it to say that a configuration in terms of control and metacontrol system does not automatically imply that one is talking in a prescriptive sense only: the same framework can be used in a descriptive sense.

APPLICATION EXAMPLE OF METADECISION-MAKING

In order to help clarify some of the concepts an illustrative example will be given with some remarks about metadecision-making. The example stems from a case study into the decision-making about 'numerus fixus' at universities in The Netherlands, that is, the procedure by which the maximum number of admissions to institutions of higher education is decided.[38] This study has centered at one particular mode of metadecision, namely the structural. One of the first things that appeared is that structural metadecision-making is almost nowhere consciously performed. One cannot pinpoint a particular body that fulfills the task of metadecision-making. The only bodies that one might interpret in this sense are the Parliament in its legislative function – because the applicable laws indeed regulate the decision-making structure – and the Capacity Committee of the Academic Council (the coordinative council of all universities) because it regularly develops proposals for the improvement of the structure of the process. Metadecision-making is rather implicit than explicit.

A general picture of the multilevel decision-making is given in Figure 7.

The environment has been omitted in this figure because of its multi-interpretability. The environmental input to the educational system consists of first year students, the output of graduates. To the universities, Academic Council or the Ministry, environment has however different interpretations. Dependent on the subsystem and the level of consideration the environment will be different.

Note that in this figure the structure that has to be decided upon, consists of the relations between the three subsystems shown (Universities, Academic Council, Ministry plus Educational System). It is however clear that inside these subsystems

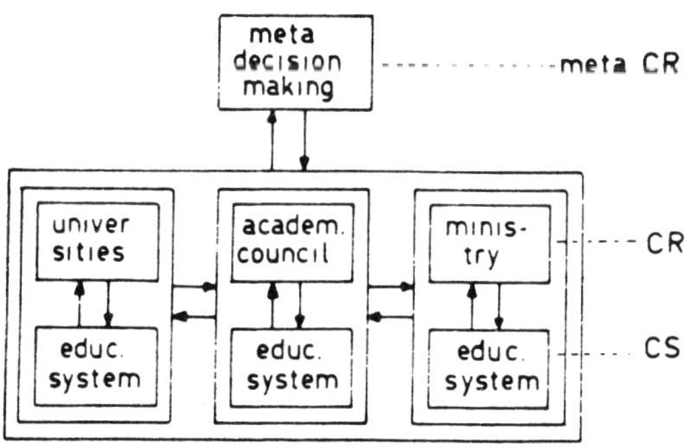

Fig. 7. The General Multilevel Decision Making System.

there also exist structures, so that there also exists metadecision-making at a lower level, for example inside a university. This structure consists of the relations between faculty council (*FR*), University council (*UR*) and University Board of Governors (*CvB*) (see Figure 8).

In an analogous way one can also subdivide the subsystem around the Academic Council into smaller parts so as to get a structure on a lower level of aggregation. The Academic Council namely consists of a plenary meeting of the Council (*AR*), a preparatory committee (*DR*) and sections per field of study. In this latter case one might even speak of a concrete metadecision-making body, namely the Capacity Committee (*CC*) (see Figure 9). Of course one can proceed with this subdivision to a lower level of aggregation, such as that of the individual members of the plenary meetings of the Academic Council (Figure 10). Also on this level there are examples of the organisation of the relations between these subsytems, that is, of metadecision-making.

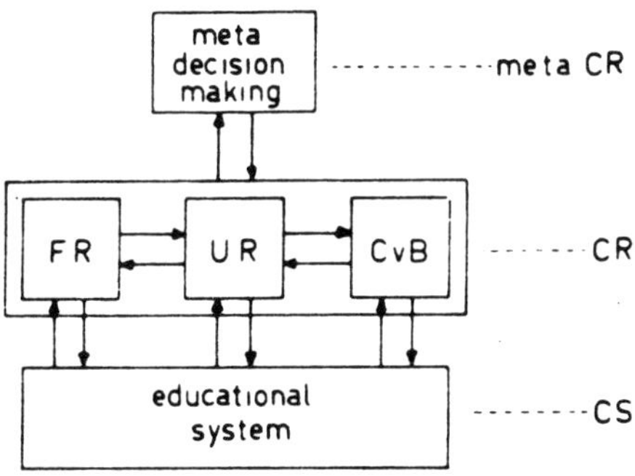

Fig. 8. The University as a Multilevel System.

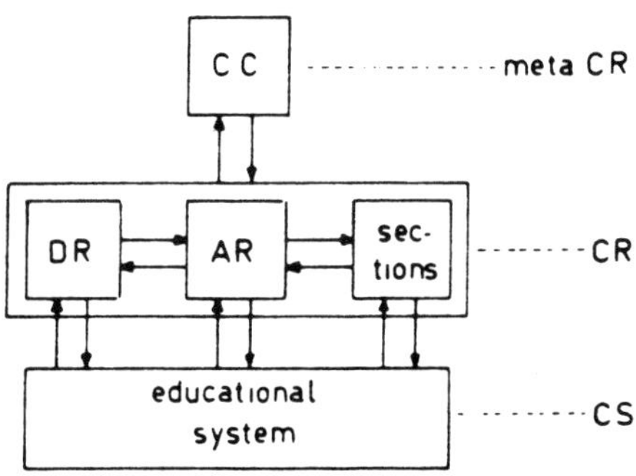

Fig. 9. The Academic Council as a Multilevel System.

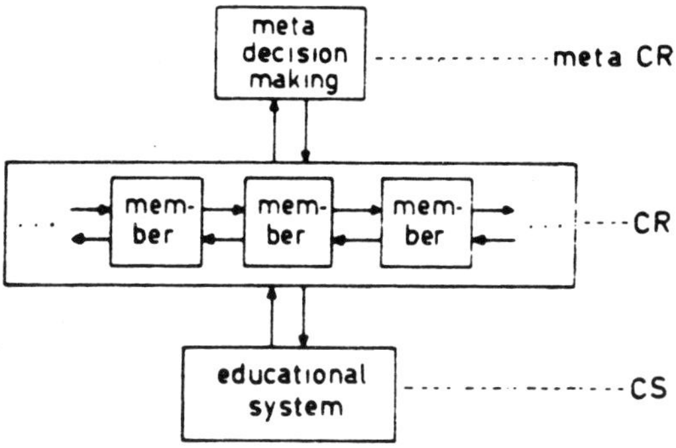

Fig. 10. The Meeting of the Academic Council as a Multilevel System.

DISCUSSION OF SOME PROBLEMS WITH METADECISION-MAKING

One of the problems we met in identifying structural metadecision-making is, for example, the question of whether changing rules inside a subsystem is also a kind of structural metadecision-making or whether this term is only reserved to change in relations between subsystems. There are two possible answers to this question. First, one can adopt the black box point of view and thus realise that changing the rules, that is changing the black box transfer function, is per definition a structural change. Second, one can subdivide the subsystem into parts so that the subsystem itself becomes a system of subsystems and relations. A change of a rule within the subsystem is therefore a change in the structure of the subsystems on the lower level of aggregation. This was illustrated in the examples of the previous section.

In broader terms this problem of rules or structure amounts to the problem of the exact definition of metadecision-making. We have introduced the concept of metadecision-making in a three level control system, with the controlled system at the lowest level, the controller at the next level and the metacontroller at the highest level. Of course it is possible to invent a multitude of different verbal descriptions of these levels such as, for example, substantive decision-making, post-policymaking, implementation, execution, and so on for the lowest level; procedural decision-making, functional decision-making, operational decision-making, and so on for the middle level; structural decision-making, goal decision-making, and so on for the metalevel. In our view, misunderstandings can be avoided by the use of the control systems point of view: define the system, the controller, the control actions, the inputs and outputs, and so on and there is no ambiguity left. One should however realise that these are not the only possible approaches and that there is no guarantee that they will be the most fruitful for developing a theory of metadecision-making. From a point of view of theoretical progress it would be better to take a theoretic-pluralism point of view.[39] Note however that our definitions are object independent and that different viewpoints yield different results. The framework possesses a large degree of freedom which actually amounts to some kind of pluralism.

The control-system description does not stop at the metasystem level. On the contrary it seems rather that one can apply an infinite iteration of meta-levels arriving at an object level, a metalevel, a meta-metalevel and so on. One can obtain this multitude of levels by continuously lowering the level of aggregation: subdividing the system into parts and relations at each level of aggregation and applying the principle of structural metadecision-making. The same can be done by looking at the goals resulting in a multilevel hierarchy of goals, subgoals, etc. (this is closely related to the well-known means-ends hierarchy). This point of view implies that the levels only differ in their amount of detail but that the aspects considered are the same. However, one can also imagine a system, metasystem and meta-metasystem configuration where the levels qualitatively differ. Take the example of an imaginary PPBS control system.[40] At the object level the PPBS is implemented, carried out. At the next higher level (the metalevel) one has to decide which kind of budgeting system should be adopted, and at the meta-metalevel there might be a body concerned with the general question of what kind of control should be applied to public administration at all.

At a still higher level one might imagine that our western cultural system plays a role. In this example there are clearly qualitative differences between the matters of concern at each different level; it is not a simple question of the amount of detail. Another example of such qualitatively different levels might be a decision-making configuration where the metalevel considers the structure of the process and the meta-metalevel considers what mode of decision-making should be adopted (rational, incremental, mixed scanning, etc.).

Maybe some light can be shed on the confusion between these two apparently different kinds of multilevel system configuration by drawing a parallel with the concept of hierarchy as defined by Simon[41]: a hierarchical system is a system which can be partitioned into subsystems, which in turn can be partitioned into sub-subsystems, and so on, to some level of elementary subsystems. This definition pictures a multilevel system constructed by iteratively subdividing the system. At first sight it would appear that the levels only differ in their amount of detail. However, this is not necessarily true, because nothing has been specified in the definition about the dimension along which the subsequent partitions take place; the aspects according to which systems are partitioned into subsystems can indeed differ from level to level. Hence a hierarchical system in which all different levels have qualitatively different matters of concern is a specific but still hierarchical system as well as a hierarchical system in which the aspects per level do not differ.

Remember, however, that the concept of a metalevel was explicitly defined as a level of qualitative difference compared to the object level. A multilevel hierarchical system, where the aspect of concern is the same at all levels, may be defined as multilevel, but does not meet the metalevel definition. The aspects per level differ in that definition.

Closely related to the problem of the difference or identity between matters of concern at the various levels is the question of whether a theory on the system level is different or isomorphic to a theory on the metasystem level. On the one hand, one might state that because we envisaged a control system approach in which the same concepts apply to a higher (meta) level, it seems probable that the theories needed to explain the phenomena at the different levels will be isomorphic too. On the other hand, it is not difficult to give a counter example in which it is clear that totally different kinds of theories are needed.

This issue is clearly related to the question of the relationship of theories on different levels of aggregation such as considered, for example, by Heiskanen.[42] Again the point is that the general concepts and ideas used in the theories at different levels may remain the same but their specific empirical contents in the particular case under consideration will, however, surely differ due to the definition of the metaconcept. Hence theories at different levels will probably be similar in a general sense but will be dissimilar in a specific empirical sense.

Another problem related to the multilevelness is the finiteness or infiniteness of the iteration of levels. As we saw in our examples one can go on talking about structural metadecision-making as long as one can subdivide a system into parts and relations. Top down one might consider some elementary level of detail (for example, the individual level) as the end of the regression just as in the earlier-mentioned definition of hierarchy. But what about a bottom up iteration? Does one stop at the level of cultural system of values or does one proceed into the universe? To answer this question would probably lead us to philosophy, something we prefer to avoid. There is, however, one thing that should be emphasized about the finiteness or infiniteness of levels, namely that in essence only two levels are considered, no matter how high the level of consideration is, that is, the level of the controlled system and the next-higher level of the controller. Even the level of metacontrol which we have called the third level, is nothing other than the application of the two-level controller-controlled system configuration at one level higher. No matter at what level one works, one always has to deal with a very finite number of levels, namely only two related levels at a time.

CONCLUSION

The intent of this paper was to draw a general conceptual framework of metadecision-making. As we have seen, problems remain. What is clearly needed is a further extension of this conceptual framework in the direction of a 'real' empirical theory of metadecision-making which can meet the scientific requirements of theories such as consistency, testability, etc. A conceptual framework can hardly be tested because of its abstract conceptual level.

However we believe that from it a coherent theory about the structural mode of metadecision-making can be derived. Explicitly incorporating the metasystem into the model of decision-making has the distinct advantage of openly setting out the roles of the system and of the metasystem; it then becomes clear that the former carries out procedures and implements standards whose design is formulated in the latter. This separation is taken for granted by most decision-makers but receives the emphasis it deserves if the decision-making model exhibits two different systems with separate functions.

REFERENCES

1. W. J. M. Kickert, *Organisation of Decision Making* (North-Holland, 1980).
2. W. J. M. Kickert, 'A Structural Theory of Organisational Decision Making.' Presented at the Fourth Int. Congress of Cybernetics and Systems, Amsterdam (August, 1978).
3. W. J. M. Kickert, *Organisation of Decision Making, Op. Cit.*
4. S. Beer, *Decision and Control* (Wiley, New York, 1966) and *Platform for Change* (Wiley, New York, 1975).
5. S. Beer, *Brain of the Firm*, 2nd Ed. (Wiley, New York, 1981).
6. C. W. Churchman, *The Design of Inquiring Systems* (Basic Books, New York, 1971).
7. G. J. Klir, 'Identification of Generative Structures in Empirical Data, *Int. J. of General Systems* **3**, 89–104 (1976).
8. Special Issue of 'Reconstructability Analysis', *Int. J. of General Systems* **7**, 1 (1981).
9. A Rapoport, 'Mathematical Aspects of General Systems Theory', *General Systems* **11**, 3–12 (1977), and *General System Theory*, (Abacus, Kent, Eng., 1986).
10. J. P. van Gigch, 'A Methodological Comparison of the Science, Systems and Metasystem Paradigms', *Int. J. Man-Machine Studies* **11**, 651–663 (1979). Reprinted in this text, Chapter 2.
11. J. P. van Gigch and L. L. Pipino, 'From Absolute, to Probable, and Fuzzy in Decision Making', *Kybernetes* **9**, 47–55 (1980).
12. W. J. M. Kickert, *Fuzzy Theories in Decision Making* (Nijhoff, Boston and Leiden, 1978).
13. J. P. van Gigch, 'A Methodological Comparison of the Science, Systems and Metasystem Paradigm', *Op. Cit.*
14. Y. Dror, *Public Policy Making Re-examined* (Chandler, Scranton, Penn., 1968).
15. Y. Dror, *Design of Policy Sciences* (Elsevier, New York 1971).
16. A. Faludi, *Planning Theory* (Pergamon, Oxford, 1973).
17. J. R. Emshoff, 'Planning the Process of Improving the Planning Process: A Case Study in Meta-Planning', *Management Science* **24**, 1095–1108 (1978).
18. I. I. Mitroff and F. Betz, 'Dialectical Decision Theory: A Meta-Theory of Decision Making', *Management Science* **19**, 11–24 (1972).
19. A. Faludi, *Planning Theory, Op. Cit.*
20. A. C. J. de Leeuw, *Organisation Science and Decision Theory*. In Dutch (Stenfert Kroese, Leiden, 1974).
21. A. C. J. de Leeuw, 'The Control Paradigm as an Aid for Understanding and Design Organisations', Third European Meeting on Cybernetics and Systems Research, Vienna (1976).
22. W. J. M. Kickert, *Organisation of Decision Making, Op. Cit.*
23. A. Faludi, *Op. Cit.*
24. Y. Dror, *Public Policy Making Re-examined, Op. Cit.*
25. A. Faludi, *Op. Cit.*
26. Y. Dror, *Public Policy Making Re-examined, Op. Cit.*
27. C. E. Lindblom, 'The Science of Muddling Through', *Public Admin. Review* **19**, 79–88 (1959).
28. A. Etzioni, 'Mixed Scanning: A "Third" Approach to Decision Making', *Public Admin. Review* **27**, 385–395 (1967).
29. A. Faludi, *Op. Cit.*
30. P. C. Nutt, 'Models for Decision Making in Organisations and Some Contextual Variables Which Stipulate Optimate Use', *Acad. Management Review* **1**, 2, 84–98 (1976).

31. J. P. van Gigch, 'A Methodological Comparison . . .', *Op. Cit.*
32. I. Heiskanen, 'Theoretical Approaches of Scientific Strategies in Administrative and Organisational Research: A Methodological Study', *Commentationes Humanarum Litterarum* **39**, 2 (Societas Scientarum Fennica, Helsinki, 1976).
33. A. C. J. de Leeuw, *Organisation Theory and System Theory*, *Op. Cit.*
34. P. Feyerabend, *Against Method* (Verso, London, 1975).
35. W. J. M. Kickert, *A Structural Theory of Organisational Decision Making*, *Op. Cit.*
36. A. C. J. de Leeuw, *Organisation Science and System Theory*, *Op. Cit.*
37. W. J. Kickert, *Organisation of Decision Making*, *Op. Cit.*
38. W. J. M. Kickert, 'Substantive and Structural Decision Making: A Case Study About the Numerus Fixus Decision Making', Int. Report. in Dutch (Eindhoven Univ. of Technology, Eindhoven, N.L. 1978).
39. P. Feyerabend, *Op. Cit.*
40. J. P. van Gigch, *Applied General Systems Theory*, 2nd Ed. (Harper and Row, New York, 1978).
41. H. A. Simon, 'The Architecture of Complexity', *Proc. Amer. Philos. Soc.* **106**, 467–482 (1962).
42. I. Heiskanen, *Op. Cit.*

4. EPISTEMOLOGICAL QUESTIONS RAISED BY THE METASYSTEM PARADIGM*

John P. van Gigch

The metasystem paradigm attempts to resolve epistemological issues which have existed from the beginning of time. These issues are recalled here to provide readers and scholars with a list to which they can refer when working on a quest for possible answers.

INTRODUCTION

Van Gigch's[1] article on a methodological comparison of the science, systems and metasystems paradigms (vide Chapter 2 in this text) raised important questions on the ability of each of these paradigms to answer epistemological issues. It showed the weakness of the science paradigm and of scientific approaches which are based on strict rationalism and reductionism. The systems paradigm was touted as the approach which would counter those weaknesses and was claimed to be able to answer the criticisms that were levelled against the science paradigm. Checkland[2,3] has expressed, better than most, the differences between the science and the system paradigms. If we date the beginning of the modern systems movement with the creation of the Society for General Systems Research in 1956, the Systems Approach has had about twenty-five to thirty years to establish itself. After this time, and for all its claims, it appears that the systems paradigm will not resolve many of the methodological issues raised against the science paradigm. As the metasystem makes its appearance on the scene with some claims of its own, it is pertinent to review these issues and evaluate whether this new paradigm can help to resolve them.

This chapter is presented in two parts. In the first part we outline a set of epistemological questions which are not addressed by the metasystem paradigm. In the second part are outlined questions which the metasystem paradigm attempts to elucidate. Each set is treated in turn.

We challenge the reader to find ways in which the metasystem paradigm or other related approach can answer some of these questions. This book is devoted to a presentation of some of the exponents of current metasystemic thinking. As is patently obvious, thinking in metasystem terms is certainly not new and is prevalent in many quarters. In order to progress, its proponents need to address themselves more explicitly to the questions left unresolved by other approaches.

Many of the chapters of this text are related to one or more of the epistemological questions raised here. We trust that, through their discussion, the various authors contribute to the elucidation of these issues.

* Reprinted and adapted with permission from: *Int. J. Man-Machine Studies*, **20**, 501–509, 1984,

SET OF EPISTEMOLOGICAL QUESTIONS WHICH HAVE NOT BEEN ADDRESSED BY THE METASYSTEM PARADIGM

The Question of Problem Definition

One of the main criticisms which the Systems Approach levelled against the science paradigm was that, when users of the latter are confronted with a real-world problem, they usually fail to properly define the problem at hand.

Invariably, they commit two sins: one, they seldom take the 'whole problem' into account, and two, their closed models do not represent accurately the world where systems are open. Users of the scientific method stand accused of reducing and fragmenting the problem into parts. To correct this flaw, the proponents of the Systems Approach advocate that a problem be expanded, that jurisdictional and geographical boundaries be pushed, in order that as many systems and metasystems be taken into account.[4] In doing so, systems scientists are faced with the problem of where to set a problem's boundaries. How far should boundaries be expanded? How many metasystems should be included? What is the optimum boundary? How is it found? Does it exist?

Checkland has tried to answer some of these questions.[5] In any problem, he defines the boundary of the 'relevant system' as (a) the system which is liable 'to solve the problem' or 'to improve the problem situation' and (b) the system which 'encapsulates the root-definition' or basic activities which form the essence of the problem under observation. Churchman[6] warned us against committing the 'environmental fallacy' which not only results when a system's environment is overlooked, but also takes place when we fail to consider the short and long term responses of outside systems upon our own. However, we cannot win because, as we try to avoid 'the environmental fallacy' and take more systems into account, we soon learn another lesson. Churchman[6] reminds us of Anaxagoras' theorem: 'In everything there is everything' or 'In every problem are to be found all problems' which means that, if a problem is unravelled far enough, we will encounter all other problems. We are thus left with the dilemma: If we push the system's boundaries too far, we face the problem of having to consider too many systems, the situation becomes too complex and unsolvable. If we do not push the boundary far enough, we face the 'environmental fallacy' when not all the relevant systems are taken into account.

The metasystem paradigm does not address itself to the so-called Error of the Third Kind[7,8] which consists of defining the wrong problem for study, neither does it help to unravel the dilemma to find the proper system boundaries or to avoid taking into account too little or too much. In Chapter 5, François raises the specter of Gödel's Theorem which has a bearing on the epistemological question of how a problem can be defined.

The Problem of Coherence or Holism

Coupled with the questions of problem definition, is the problem of coherence or holism. This is concerned with the need propounded by the systems paradigm to study 'irreducible wholes' which contain all the attributes of the system. In our urge to avoid reductionism and the environmental fallacy, the boundaries of the 'relevant system' are to be pushed to encompass more subsystems. When are the boundaries of this 'coherent whole' reached? As a collateral but related question, which world view should we adopt: the atomistic reductionist view or the holistic global view?

Bunge[9,10] is prompt to argue that true systems theory can 'neither be atomistic nor holistic'. It cannot completely reject the atomistic approach, the atomistic ontology and the atomistic methodology of the traditional methods of science, because it must proceed by analysis to study the way components act upon one another. Systems theory cannot be merely holistic if, by this term, is meant an approach which contends that knowledge of a system only consists of an understanding of the emergent properties and behaviour of the totality. Indeed, systems theory must accept the compatibility of the atomistic and holistic approach, because it probably overlaps both. Given that the metasystem approach can be considered a general systems theory, it should neither be strictly atomistic or holistic, but amenable to both approaches.

The Question of Complexity and Solvability

As we push the systems' boundaries to attain the relevant system, avoid the Error of the Third Kind and define the coherent whole to deal with the total system, we fall into another dilemma: the one which besets those whose system's complexity overwhelms the decision-maker's capacity to deal with it. As we expand a system's horizon, the domain becomes more complex and its solvability is reduced. Thus as we struggle to take more systems into account we diminish our chances to find the problem's solution. We totter between *simplicity* and *complexity* and between *solvability* and *incomprehension*. Does the introduction of the metasystem approach help or compound the problem? It is probably an impossible question to answer definitely. To the extent that the metasystem paradigm has contributed to a better understanding of the interaction of the various levels of control in a systems hierarchy, it contributes to the side of solvability and simplification. The application of the paradigm to the study of dysfunctions and failures[11–14] provides indications of progress in the direction of unravelling and controlling complexity.

The Question of Idealism and Realism

Churchman[15] outlined the dilemma between the idealist and the realist. The latter has the best technology, the former the best philosophy. The realist is the type of planner who adopts approximate, nearby, solutions and implements them in a practical way. The idealist is the planner who champions optimum solutions that are never implemented. In a way, the question of problem definition leads into the realist–idealist dilemma, because the realist represents the scientist who refuses to push a problem's boundary to the point at which it becomes unsolvable – the point where, the idealist claims, the best solution lies.

It can be claimed that the metasystem paradigm adds both to the world view (philosophy) of the idealist and to the methods (technology) employed by the realist. However, these claims are premature because the paradigm has only recently been applied to the systems domain.

SET OF EPISTEMOLOGICAL QUESTIONS ADDRESSED BY THE METASYSTEM PARADIGM

The Problem of Optimization and Suboptimization

The idealist–realist question raised earlier, leads into the optimization–suboptimization dilemma. To attempt the ideal solution is to attempt the futile target of optimization. At best, we suboptimize, which is the realist's solution. How does this indictment apply to the process of decision-making in real-world organizations? Obviously, Simon's famous motto[16] that decision-makers 'do not have the wit to maximize and can, at best, only satisfy' comes to mind. 'To satisfy' is the embodiment of the suboptimizing stance of the realist. Given the boundaries of rationality of the manager, optimization only exists in the confines of exact models. How does the metasystem paradigm contribute to this debate?

The metasystem paradigm leads to a representation of the organization as a hierarchy of levels and of decision-making as a control process carried out at several levels of recursion.[17,18] Attempts to formalize the mathematical theory of optimization in a multilevel hierarchy continue unabated, as evidenced by Basar.[19] However, it is in the mapping of the flow of information and of rationalities among levels, and in the improved understanding of the decision process in complex multilevel organizations that the metasystem approach shows its strength. Van Gigch[20] and van Gigch *et al.*[21] discuss the decision process in actual situations. These authors show how the operational, tactical and strategic levels interrelate, and how the rationalities of one level serve to make decisions at another and thus are named metarationalities. Once the idealistic premise of optimization is discarded, it is important to settle for a more realistic view of how organizational decision-making is reached. The metasystem paradigm also serves to characterize the problem of determining the optimum dosage of centralization–decentralization suitable to each organization.

The Principle of Invariance

General Systems Theory was premised on the existence of isomorphisms which could be found among systems of different domains and disciplines as well as different levels of a hierarchy.

Laszlo[22] argued for invariance across systems levels and explained how systems evolve towards higher complexity through quantum jumps which generate the stable configurations of organization levels.

That isomorphism reigns in the hierarchical constructs of the metasystem paradigm is debatable.[23,24] As a matter of fact, one of the characteristics of the metasystem paradigm is its emphasis on the fact that there is *no isomorphism* of problem domain and of rationalities across the various levels of organizational decision-making. The evidence resulting from the application of the metasystem paradigm to organizations shows that each hierarchical level deals with a *different problem* and that the epistemology, methodology and rationality required to solve each of the problems is clearly distinct and *different from one level to another*. This distinction and lack of isomorphism places new importance on rethinking how complexity of hierarchical systems has evolved. Invariance does not rule out the evolutionary process, but must take into account differentiation.

The Ontological Question

Do hierarchies really exist or are they figments of our imaginations – a practical way to model organizational decision-making? Regardless of where the truth lies, the metasystem paradigm's very existence rests on the premise that an organization *is* a hierarchy made up of levels of control systems.

In applications of the metasystem paradigm to management problems, its prescriptive value has been tested.[25,26] The reach of these prescriptions is rather meagre to date. The metasystem paradigm works better as a *descriptive* than as a *prescriptive* one. Locker and Coulter[27] and Kickert[28] discussed the interplay of descriptive and prescriptive models in the context of the metasystem approach. Troncale[29] discussed the role of hierarchies in science and metascience.

The Paradigm as a Guarantor of Truth

Van Gigch[30] showed the difficulties faced by the science and the systems paradigm in the evaluation and proof of truth. In the science paradigm, it is usual to formulate objective functions in terms of measures of effectiveness which embody the models or decision-maker's goals. The Operations Research model, a component of the science paradigm, is 'self-serving' or tautological in that it provides in the model, the function by which the solution can be tested for optimality. Apart from attempts to evaluate the correspondence between model and reality and to offer improvements between the old and new systems, it does not offer a guarantor outside the model.

There are several versions of the systems paradigm which can be reviewed to assess how they are used to evaluate truth. For Checkland[31], changes to the system must meet two criteria. They must be *desirable* and *feasible*. The more comprehensive approach advocated by van Gigch[32] favours the use of multidimensional decision models where more than two criteria can be satisfied. He calls for compromise and tradeoffs, and attempts to implement consensus. He also calls for the incorporation of ethical considerations in the calculus of systems optimization. In Churchman's[33] terms, systems science must be 'self reflective' and evaluate the results of its own deeds. We will return to the ethical issue later, to see how it is dealt with by the metasystem paradigm.

Concerning the proof of truth, Checkland[34] contends that the choice of a desirable change stems from the comparison between the conceptual model of the 'root definition' and descriptive model of a perceived reality. Here, consensus provides the guarantor of truth with assurance that, (a) the descriptive model reflects the real world, (b) the root definition contains the essential activities of the 'relevant system' and, (c) a set of desirable and feasible changes to affect reconciliation between model and root definition can be agreed upon. Consensus is also invoked by van Gigch[35] as a necessary ingredient of truth, but its practical validity may be illusionary when dealing at the scale of the global system. Nadler[36] offers a time-line perspective which, like Checkland, compares how the system *is* with what it *ought to be*. Thus, none of the exponents of the system paradigm can be said to resolve the question of truth evaluation and proof.

It is to fill this void that the metasystem paradigm was created. As stated in Chapter 2, it incorporates:

(1) a hierarchy of problem solving levels in which higher systems level can judge and rate solutions for lower ones;
(2) a framework to provide evaluation criteria in metalanguage terms, that is, in a language appropriate to judge lower system solutions;
(3) a guarantor of truth at each system level, but the last (or highest).

Beer[37–39] introduced all of those characteristics in his formulation of the metasystem in organizations. Metasystem criteria are always needed to judge 'object' systems criteria. Lower level decisions can only be arbitrated in *metalanguage* terms. The question of metalanguages is taken up by Nowakowska in Chapter 8.

The metasystem must evaluate how object level systems apply technology to produce waste as well as wealth. The metasystem is also in a position to evaluate the ethical implications of systems designs proposed by lower levels, an issue embodied in the Morality of Systems.[40]

The idea of multi-level ethics was conceptualized by Jantsch[41,42] and Zeleny.[43] Jantsch did not specifically mention the metasystem paradigm, but conceptualized morality as the *aggregated evaluative rationality* which guides decisions at each level of the organization. It is important to note that this rationality is dynamic and that the criteria which represent it at each level can be in conflict with one another, that is, there is a different perception of morality at each level. Furthermore, Jantsch introduced the novel thought that the morality or evaluative rationality of the organization's metapolicy level must act as a *variety amplifier* rather than as a *variety attenuator*. It must be open to the environment to ensure the organization's survival. Lower levels' morality must evolve and keep in step with the morality of the metasystem. The whole concept that higher system levels can prescribe for lower ones is contested when Gödel's Theorem is invoked. See François in Chapter 5 in this text.

The Question of Observability and Control

The rationalism background of the scientific method and of the science paradigm ensures that they be wedded to the epistemological concept of object/subject split, to the myth of objectivity and to the notion of neutrality and non-involvement of science. None of these principles can hold either in the context of the systems paradigm or of the metasystem paradigm.

The systems paradigm as well as the metasystem paradigm are based on the premise of the indivisibility of domains of organized complexity, where entities are characterized by openness, low separability and high interdependence of the parts. These paradigms also reject the notion that the observer/designer can separate itself from the observed system and that the subject/object split is real. As Locker and Coulter[44] have shown, there is a constant interplay between observer and designer and between description and prescription.

Coulter[45,46] and Coulter and Johnson[47] reminded us that any alleged property of the system being observed pertains to the perspective of the observer and that the role of purpose is crucial to the metasystem paradigm and to teleogenetic systems. How do teleogenetic systems guarantee truth? They can generate their own goals at all levels, including the metapolicy level. Hence, evaluation of actions and results against goals is also possible at the metalevel. Thus it would appear that teleogenetic systems have a built-in guarantor.

In the relativistic world of Jumarie,[48,49] a system's finality is defined by the observer. There is no system without an observer, no two observers are identical, and every observer is subjectively selective. Observed system and observer depend on the mutual and relativistic definition of one another. Man cannot evaluate his own abstract systems in the abstract. A Universe *U* is blind, in the sense that it cannot observe its own variations and changes. Consequently, conscious human beings are witnesses and observers of systems, but the only systems about which they can obtain knowledge are the systems they observe.

CONCLUSION

The epistemological issues discussed above have been with us from time immemorial and it would be presumptuous to think that they will be resolved by any new logic very soon. It is useful to bring them forth to our attention to mount a concerted effort which may lead to new solution pathways. The metasystem paradigm is touted as one of these approaches. It is up to the reader to pick the challenge up and to continue this quest.

REFERENCES

1. J. P. van Gigch, 'A Methodological Comparison of the Science, Systems and Metasystem Paradigms', *Int. J. Man-Machine Studies* **11**, 651–663 (1979). Reprinted as Chapter 2 in this text.
2. P. B. Checkland, 'Science and the Systems Paradigm', *Int. Journal of General Systems* **3**, 127–134 (1976).
3. P. B. Checkland, *Systems Thinking, Systems Practice* (Chichester, England and New York, Wiley, 1981).
4. J. P. van Gigch, *Applied General Systems Theory*, 2nd Ed. (New York and London, Harper and Row, 1978).
5. P. B. Checkland, see references 2 & 3, *Op. Cit.*

6. C. W. Churchman, *The Systems Approach and Its Enemies* (New York, Basic Books, 1979).
7. I. I. Mitroff and T. M. Featheringham, 'On Systemic Problem Solving and the Error of the Third Kind', *Behavioural Science* **19**, 383–393 (1974). I. I. Mitroff, 'Toward a Theory of Systemic Problem Solving: Prospects and Paradoxes', *Journal of General Systems* **4**, 47–59 (1977).
8. R. H. Kilmann and I. I. Mitroff, 'Problem Defining and the Consulting/Interview Process', *California Management Review* **21**, 26–33 (Spring 1979).
9. M. Bunge, 'General Systems and Holism', *General Systems* **22**, 87–90 (1977).
10. M. Bunge, 'The GST Challenge to the Classical Philosophies of Science', *Int. Journal of General Systems* **4**, 27–29 (1977).
11. J. P. van Gigch and W. J. M. Kickert, 'Applications of the Metasystem Paradigm to Organizational Decision Making', *Applied Systems and Cybernetics, Proceedings of the International Congress on Applied Systems Research and Cybernetics*, Acapulco, December 1980, Edited by G. E. Lasker (New York, Pergamon Press, 1981).
12. J. P. van Gigch, 'The Metasystem Approach to Systems Failures', *Systems Science and Science, Proceedings of the 26th Annual Meeting of the Society for General Systems Research*, Washington, D.C. (January, 1982).
13. J. P. van Gigch, 'Toward Descriptive and Prescriptive Theories of Systems Failures', *Proceedings of the 27th Annual Meeting of the Society for General Systems Research*, Detroit, Michigan (May, 1983).
14. A. F. van Gigch, 'The Application of the Metasystem Paradigm to Cost Containment in a Health Care Program' (Vide Chapter 20 in this text).
15. C. W. Churchman, *Challenge to Reason* (New York, McGraw-Hill, 1968).
16. H. A. Simon, *Administrative Behaviour*, 3rd Ed. (New York, Free Press, 1976).
17. S. Beer, *The Heart of Enterprise* (New York, Wiley, 1979).
18. W. J. M. Kickert and J. P. van Gigch, 'A Metasystem Approach to Organizational Decision Making', *Management Science* **25**, 1217–1231 (1979). Reprinted as Chapter 3 in this text.
19. T. Basar, 'Stochastic Multicriteria Decision Problems with Multi-levels of Hierarchy', *IEEE Transactions on Automatic Control* AC–26, 549–553 (1981).
20. J. P. van Gigch. See references 11–13, *Op. Cit.* (See also Chapter 20 and 21 in this text.)
21. J. P. van Gigch *et al.*, 'Applications of the Metasystem Paradigm to the Diagnosis of Systems Malfunctions'. Student Projects presented in connection with a *Seminar in Systems Analysis*, California State University, Sacramento, School of Business and Public Administration, Several Semesters (1981–1984).
22. E. Laszlo, 'A General Systems View of Evolution and Invariance', *General Systems* **9**, 37–43 (1974). E. Laszlo and H. Margenau, 'The Emergence of Integrative Concepts in Contemporary Science', *Philosophy of Science* **39**, 25–259 (1972).
23. W. J. M. Kickert and J. P. van Gigch. See reference 18, *Op. Cit.*
24. J. P. van Gigch. See Chapters 18 and 19 in this text.
25. J. P. van Gigch *et al.* See reference 21, *Op. Cit.*
26. A. F. van Gigch. See reference 14, *Op. Cit.*
27. A. Locker and N. A. Coulter Jr., 'A New Look at the Description and Prescription of Systems', *Behavioural Science* **20**, 179–190 (1975).
28. W. J. M. Kickert, *Organization of Decision Making* (Amsterdam, North Holland, 1980).
29. L. R. Troncale, 'Science, Metascience, Metaphysics: A Spectrum of Ways of Knowing' (Abstract), *Systems Science and Science, Proceedings of the 26th Annual SGSR Meeting*, Washington, D.C. (January, 1982).
30. J. P. van Gigch, 1979, *Op. Cit.* reference 1.
31. P. B. Checkland, 1976, *Op. Cit.* reference 2.
32. J. P. van Gigch, 1978, *Op. Cit.* reference 4.
33. C. W. Churchman, 1968, *Op. Cit.* reference 15.

34. P. B. Checkland, 1981, *Op. Cit.* reference 3.
35. J. P. van Gigch, 1978, *Op. Cit.* reference 4.
36. G. Nadler, 'A Systems Engineering Approach to Securing Real World Changes: A Timeline Perspective', *J. Applied Systems Analysis* **6**, 89–100 (1979).
37. S. Beer, *Brain of the Firm* (New York, Wiley, 2nd Ed., 1981).
38. S. Beer, *Platform for Change* (New York, Wiley, 1975).
39. S. Beer, *The Heart of Enterprise* (New York, Wiley, 1979)
40. C. W. Churchman, 1968, *Op. Cit.* and J. P. van Gigch, 1978, *Op. Cit.* references 15 and 4, respectively.
41. E. Jantsch, *Design for Evolution* (New York, Braziller, 1975).
42. E. Jantsch, *The Self-Organizing Universe* (Oxford and New York, Pergamon, 1980).
43. M. Zeleny, *Autopoiesis: A Theory of Living Organization* (New York and Oxford, North Holland, 1981).
44. A. Locker and N. A. Coulter Jr., *Op. Cit.* reference 27.
45. N. A. Coulter Jr., 'The Self-Determinism of Teleogenetic Systems', *J. Cybernetics* **5**, 9–20 (1976).
46. N. A. Coulter Jr., *Synergetics: An Adventure in Human Development* (Englewood Cliffs, New Jersey, Prentice Hall, 1976).
47. N. A. Coulter Jr. and A. L. Johnson, 'Teleogenetic Systems Theory as a Metasystem Methodology', *Systems Science and Science, Proceedings of the 26th Annual SGSR Meeting*, Washington D.C. (January, 1982). Chapter 12 in this text.
48. G. Jumarie, 'The Concept of Structural Entropy and Its Applications to General Systems', *Int. J. General Systems* **5**, 99–120 (1979).
49. G. Jumarie, *Subjectivité, Information, Système-Synthèse pour une Cybernétique Relativiste* (Ed. Universe, Montréal, 1980).

5. CONSEQUENCES OF GÖDEL'S THEOREM FOR THE METASYSTEM PARADIGM

C.O. François

Given that the metasystem paradigm postulates the need to resort to a higher level of logic to guarantee a lower truth, it is pertinent to raise the specter of Gödel's Theorem, which negates, on axiomatical and mathematical grounds, that such a guarantee is possible. Gödel's Theorem asserts that all generally used logical and consistent systems are incomplete. There always exist true statements expressible in the language of a particular system, which cannot be proven in the logic of the system itself. One must always resort to the logic of an outside system to establish a criterion of its validity.

SYSTEMS AND METASYSTEMS RIDDLES IN LOGICAL TERMS

The metasystem paradigm has many implications not clearly understandable without reference to various aspects of modern logic, all of them somehow interconnected. The idea of the present paper is not to explore these interconnections which would be proper for a meeting of logicians and mathematicians. Whatever may be the coherences or contradictions, apparent or real, between these concepts, the interesting point for us, is to reconsider the metasystemic paradigm with reference to its possible logical justification and to the implicit rationale of human reasoning.

Thus those different concepts related to the supposed capacity of man to understand his own way of thinking are only to be enunciated and tentatively interconnected, in a very simplified way.

Gödel's theorem states basically that no formal system can be fully justified from within itself, but that on the contrary there is an unavoidable need to use more powerful proofs from outside the system.

Following Alonzo Church, any formal, solvable and not contradictory system contains necessarily undecidable propositions.[1]

Both theorems are obviously related. But they seem to tell us something else. For example, the self-reference calculus of Spencer Brown[2,3] introduced a kind of circularity in reasoning and would allow for self-reproduction in a *closed system*.

Do systems of that kind really exist? Is it possible to enclose a system totally within such conceptual brackets? Is that an implicit contradiction to the metasystem paradigm? If not, how do we reconcile self reproduction in close circle with metasystemic dependence?

In a more general systemic language: how do we connect homeostasis with emergence? Is evolution compatible with self-reference? How can noise produce order if not by change?

And if we use cybernetic language: may controls be totally endogenous to a system? If not, how does a metacontrol become internalised within a system? What is a metacontrol and how does it work?

Should we look for practical lessons in the management of human affairs, we must still add the following questions: has the highest value system a justifiable metasystem? Where is the outermost of these Russian dolls?

SYSTEMS AND METASYSTEMS RIDDLES IN GENERAL SYSTEMS TERMS

The Control of Contradiction

Cybernetic control has been described many times. Its mechanisms are well known and it has been thoroughly modelled. A great number of examples of controls has been produced. However, the ultimate meaning of control is possibly not so clear, nor how it appears in natural systems (not man-made or man-managed).

Control needs negative or positive feedback to start with, and later, an appropriate alternance of negative and positive feedback, as to produce homeostasis (perfect or imperfect) or asymptotic stability.

If we try to understand the genesis of a control, what we observe is a system growing apace within its significative environment. After some time, the system starts to impinge more and more strongly on its environment and, later, begins to suffer a progressive degradation. Being unable to feed without limits, the growing system produces a negative feedback effect and the growth of the system may decline and even be destroyed. If the pressure of the host is sufficiently lessened, the environment recuperates its bearing capacity and the system may grow again or become stabilized.

Should such an interrelation not exist or should it appear too late, the system would explode, which means that it would use up the totality of the resources present in its environment. For lack of limits, it would burn itself out.

To avoid this end, the interaction between the system and its environment must become steady. This implies a succession of negative and positive feedbacks of the environment upon the system and vice-versa which represents, basically, a control of the system by its metasystem. But the relation is obviously reciprocal, be it parasitic or symbiotic: the control itself is a dual interrelation that cannot be dismembered.

The control is most generally cyclical. However, this is sometimes difficult to see, because numerous different controls do interfere and create seemingly irregular movements. In these cyclical rhythms there seems to be a kind of endogenous resonance to the exogenous control. Some inside subsystem, endowed with the necessary variety (as described by Ashby) takes charge of control. It becomes then the classical control of outside variations, following the law of requisite variety: homoeostasis is instituted in the system.

This endogenous control is obviously based on the continual reconstruction of the elements of the system and of their interrelations. It is done by way of numerous cyclical processes also based on positive and negative feedbacks. To obtain such a result which is conducive to dynamic stability, a kind of basic pattern, must be constantly reproduced.

This auto-reproduction seems to be of a catalytic nature[3,4] or cyclically self-referent, based on intermediate elements that serve as auxiliaries in a series of transactive reactions.

In this way, we have crossed the chasm between exogenous control by the metasystem and endogenous autonomous control by self-reproduction.

Or have we? Is self reference really possible independently from outside interference? And if so, under what conditions and limits? On the other hand, would homeostasis be possible without that precise kind of systemic self reference?

Control for Stability

Let us try to unravel the knot in a different way.

If self-reference is really a property of closed systems, then it becomes very difficult to understand how evolution is possible. Varela, for example, insists that organization is something quite different from a particular structure. Organization should be a general and unique principle, the source of every structure, at least in living systems. This could be admitted if there is a way to explain the progressive appearance of more and more complex structures. Any individual autonomous and homeostatic system must necessarily be self-reproductive in order to avoid self-destruction. It exists generally as a mere element in a vast pool of other systems of the same type, *but none identical to any other.*

If we observe populations, or quasi-systems (in Eigen's meaning) or scattered systems,[5] we see that something quite different obtains: statistical stability through fluctuations with a threshold constraint.

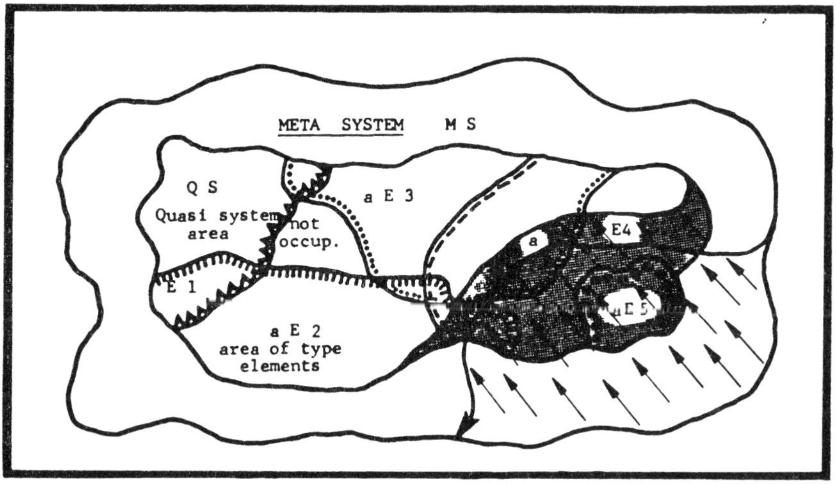

Fig. 1. Fluctuating Stability of Quasi Systems.
Ground lost by the QS because of a reduction of the MS.
An E group of any type can be wiped out only by annihilation of all its elements, by direct action of the MS on an area of the QS.
Examples: advance of an ice sheet, or a desert.

Stability results admittedly from competition between selectively equivalent replicative units that preserve (under adequate coefficient) the global information of the system (Eigen).[4,5] However, we must ask: From where do these 'adequate' coefficients originate and what is the meaning of 'adequate'? Does 'adequate' mean what insures the survival of the quasi-system under certain defined conditions?

Populations' (or scattered systems') dynamic stability in a more or less cyclically changing environment is possible only by the progressive and shifting replacement of subtypes. Every subtype is represented by a variable number of elements or individuals. These individuals are self-reproductive, and they remain true to type during their whole lifetime.

We observe thus a curious situation. The quasi-system maintains its stability within its metasystem environment, by way of a lax control of the same environment upon the elements of the system (see Figure 1). Each of these elements is endowed with definite and stable characteristics that allow or do not allow it to survive. The quasi-system is fuzzily organized, but each of its components is strictly determined – within limits of tolerance — and directly controlled by the metasystem.

This is the static view. We can understand what the situation is, but to understand how it arose is quite another matter.

Self-reference and Self-organization

We should start perhaps with self-organization (or autogenesis) as opposed to self-replication. As gatherers of information, organized systems are ambiguous in relation with the second principle of thermodynamics: being organizers and organized, they constitute local zones of negative entropy, but, at the same time, the price must be paid by an accelerated entropization of the system-metasystem sum.

The world of living systems and possibly the whole physical universe thus appears as a device for reconcentration of energy, whose activity through time, corresponds with evolution.

We generally perceive systemic structuring as a *progressive* superposition of self-organising layers. This is for example very clear in Howard Odum's[6] work, a noted exponent of ecological thinking. Trophic levels correspond to complexity, but they must necessarily have appeared diachronically (see Figure 2).

Thus, we have a problem of emergence, or of threshold, that seems precisely to be related to the growing competition of various systems within an environment, acting as a global metasystem, absorbed in a progressive way by the new and higher levels of organization within the evolutive setting.

Two general theories are presently being elaborated in relation with this threshold crossing. Both present very intriguing features and both seem very important as generators of General Systems models.

The first one derives from the concept of self-organization in the non-equilibrium systems of I. Prigogine and his school of new thermodynamics which describes 'order through fluctuations' and the appearance of 'dissipative structures'.[7]

In this work, the difference lies between a steady state system, dynamically

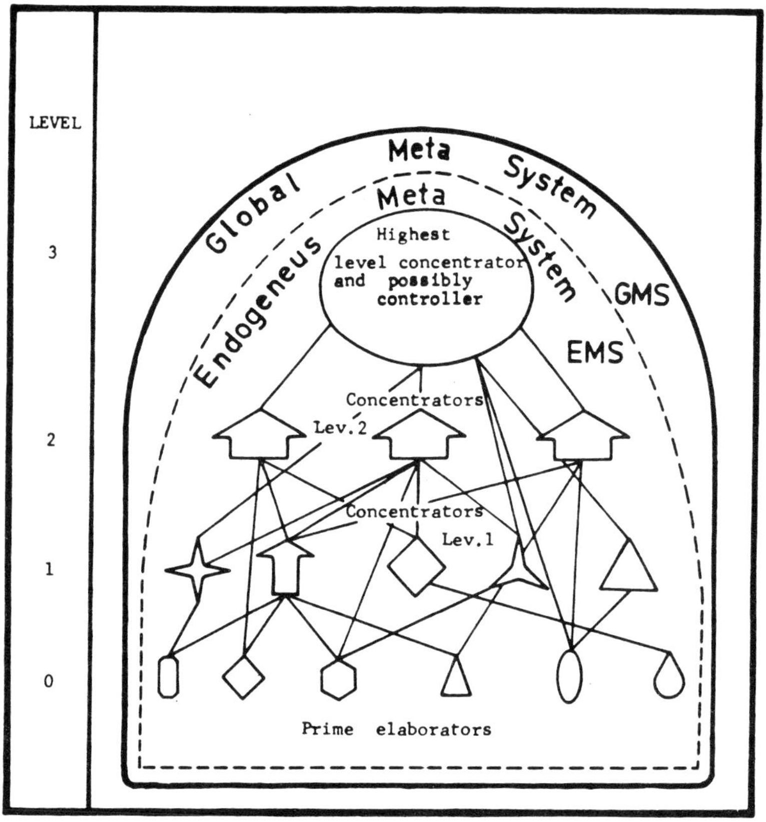

Fig. 2. Hierarchized Organization.

stable within a vicinity and autoreferent within such limits, and an evolutive system that tends to a higher complexity by elaboration of more differentiated structures. At the end of their book, the authors ask the crucial question 'whether there is a limit to complexity'. This could be formulated in alternate ways, as for example:

- Is there a limit to the progressive encroachment of pyramidal growth of organization on the global metasystem?
- What is the difference between the informal subordination to a less organized but energy-rich metasystem and the endogenously grown metasystem whose control tends to dynamic stability?

Note that the endogenously generated metasystem is still dependent on the global one.

In any case, Nicolis and Prigogine conclude that 'a sufficiently complex system is generally in a metastable state'. They add: the value of the threshold for metastability depends in a complicated fashion on the system's parameters *and the external conditions*' (this author's emphasis).

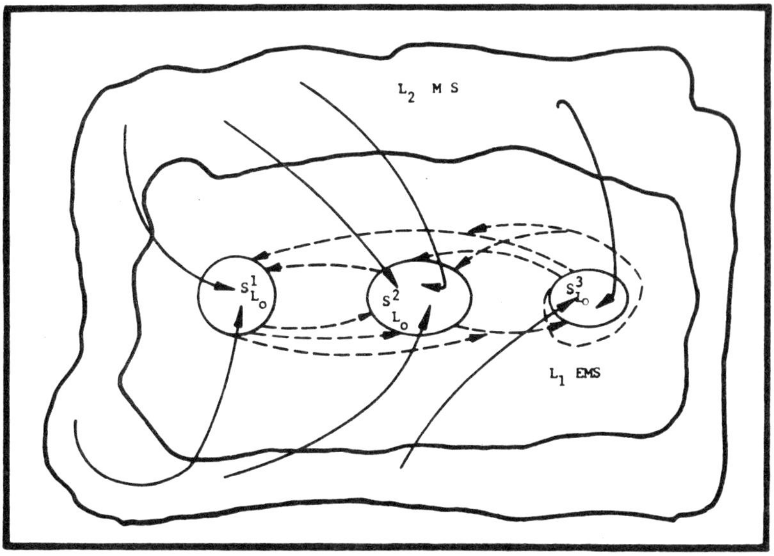

Fig. 3. Genesis of the Endogenous Meta System.
The endogenous metasystem on level 1(L_1EMS) is generated by growing reciprocal control between various level 0 systems (S_{L_0}) directly feeded by inputs from the meta system (L_2 MS).

Our study leads us finally to distinguish between an exogenous Global Meta System, energy provider to basic level systems and an endogenous organizational metasystem grown out of energy dissipation within this lower level system (Figure 3).

This concept is not contradictory with the second theory which is presently being elaborated in connection with threshold crossing. This second theory belongs to Eigen and Schuster's Hypercycle[8]. These authors also distinguish between stable systems, characterized by stability of information, under specific conditions, and evolution, related to error thresholds. In rare cases, equivalent to Prigogine's 'order through fluctuations', ring closure in dynamic systems leads to entirely new properties of the whole system. It leads to the hyper-cycle, a metastructure controlling the incorporated and interconnected systems. The hypercycle re-establishes dynamic stability (and self-reference) *at a higher level* and through some kind of innovative jump.

The difficult problem of relating in a satisfactory way Varela's self-reference, Prigogine's self-organization and Eigen's hypercycle will not be discussed here.

METASYSTEMS RECONSIDERED

From Gödel and Church to Prigogine and Eigen; from Spencer Brown to Varela, a broad spectrum of concepts have appeared in the last fifty years, nearly simultaneously, in widely different fields of science.

Van Gigch's GST[9] and his concept of metasystem are very closely related to these concepts and thus explain the above comments. Van Gigch applies the metasystem concept to the study of human organizations and, more precisely, to the levels of their control. It is thus important to try to re-evaluate it and to consider its implications.

Outer and Inner Metasystems

Human socieities in general, and human groups and organizations in particular, are dependent upon an ecological metasystem of variable type and size, to determine their structure and organization. (See Figure 4). This ecological metasystem is the provider of energy to the human system. The latter could not exist without this basic input.

However the size and nature of the metasystem depend on the information elaborated by the human group. Oil, for example, was included in the ecological metasystem of man during the second half of 19th century, *after its possible uses were understood.*

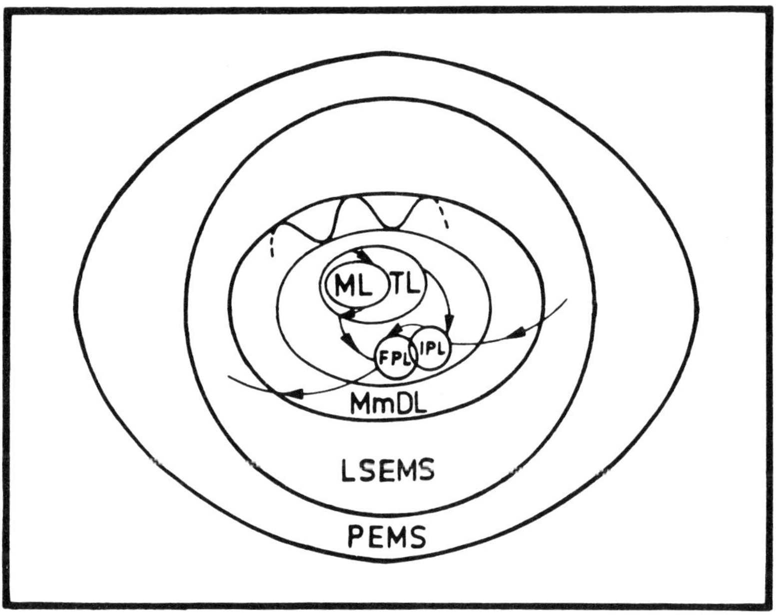

Fig. 4. Basic Structural Model of Homeostatic Human Systems.

PEMS Planetary Eco-Metasystem.
LSEMS Local or Specific Eco-Metasystem.
MmDL Max/minimum Demographic Level.
ML Mental Level.
FPL Food Production Level.
IPL Implements Prod. Level.
TL Technical Level.

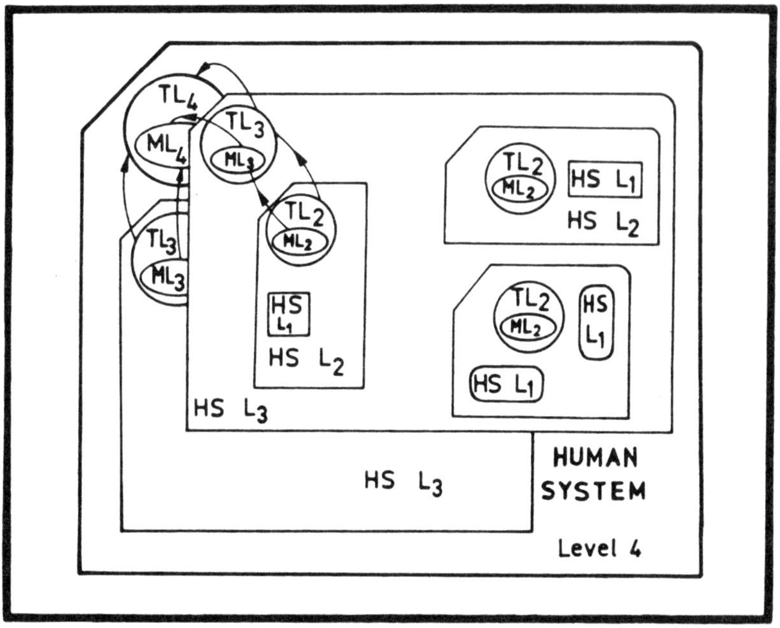

Fig. 5. Evolutive Dynamics of Human Systems.
–Crossing mental and technical thresholds.
–Assimilating the elements of lower level systems. liberated from destroyed sub-systems.

This is a general case of extraordinary importance. The evolution towards more complex systems is marked by progressive emergence of control of small systems by larger ones in the making.

Human systems, starting with the most elemental, used the inputs received from the ecological metasystem to build their own metasystems, following the general model of Figure 3 (see Figure 5).

In this way, we have two different kinds of controlling metasystems: the human and the ecological. They are interrelated, given that the second could not exist without the first one. But somehow, the human metasystem is gaining a growing control over the ecological metasystem. These little understood feedbacks should be investigated more closely: it is of paramount importance for the future of mankind.

Metasystems are not Predictable

Human metasystems construct themselves. At the present moment we are just beginning to understand that we are controlled by our own metasystem and undoubtedly it is a very startling discovery, whose potential consequences are anybody's guess. Because, as individual actors, we are very puny elements in our metasystems, we perceive them a long time after their spontaneous emergence. To be more correct, a very small percentage of people do so, while others are unaware of the invisible controls to which they are submitted. Moreover, when we try to know what really happens to us, we discover that we do not even understand how it all came about. How were our metasystems built, is still a baffling mystery. Nobody in the past seems to have been able to predict our present. The logic inscribed within the metasystem appears as sets of laws, norms and codes. Generally, we obey them, but we do not know why they are what they are.

Metasystems are metadeciders for us. As van Gigch implies, we can climb the ladder as high as possible in order to try to understand what a decision is. But we finally discover that the uppermost deciders are also decided, just as in the fable of the dictator that had full confidence in his prime minister that was controlled by his analyst that was under the spell of the dictator's rabble-rouser and so on.

The future evolution of our metasystems is at present totally unpredictable. Possibly the only way to overcome this ignorance is to study them along the lines proposed by Gödel, Church, Prigogine, Eigen and Varela. From general laws of metasystems genesis, we may be able to obtain some more precise ideas about our own socio-cultural metasystems.

The (meta) System Without a Metasystem

Van Gigch's ladder is bound to provoke a powerful vertigo in us as we reach the uppermost rung and discover that we are high in thin air, without anything in our reach to grasp or to lean on. When we climb to the level of the highest deciders, we find out that they act in accordance to a set of values which, in the best of cases, are universally recognised as undisputable.

From where did the supreme metasystem emerge? And where is the metasystem that may explain it? In curious agreement with Gödel's theorem, it does not seem possible to justify it from within the system.

Let us observe a contemporary case of value emergence.

The Christian concept of man's relation with his environment was until recently fairly well summarized in the idea that God had given the Earth to Man as his property. Western culture came even to believe that God's mandate gave Man the right to exploit Earth in whatever way he saw fit for his immediate benefit. The environmental metasystem allowed for much leeway as long as it was not too hard pressed. However, during the last twenty years, growing apace, Mankind started to live under the menace of its own abuses.

Emergence of an ecological conscience in recent years is a case of newly-born metasystemic values. These values surge in the mind of individuals as a direct effect of the impact of the ecological metasystem on the elements of the system (see Figure 3). Novice individual actions provoked global disturbances in the environment. The feedback is again felt by a few individuals first, by many more later. It has obviously to do with the growing *density* of the human systems. Their impact is more powerful when people are more numerous (the perverse effects of Raymond Boudon[11]).

Later another and very curious consequence of that higher density shows up: the closer contact between people generates also a global effect and produces a consensus on the new values. In this way, we have a closing circle, a kind of socio-cultural catalysis quite similar to Eigen's hypercycle model.

When the consensus becomes very general and starts to be transmitted, nobody remembers the cultural mutants who started the new way of thinking. We have now a modified metasystem that is starting to perpetuate itself by self-reference[12] and we are back to dynamic stability.

The ultimate endogenous metasystem is autogenerated and becomes automatically stabilized. However, it is impossible to explain its genesis without a triple reference to the environmental metasystem, to its action on the subsystemic elements and to the densification of the latter. Maybe, after all, we are on the verge of predictability of socio-cultural metasystem mutations.

REFERENCES

1. J. Ladrière, *Les limitations internes des formalismes: Etudes sur la Signification du Théorem de Gödel* (Nauwelaerts, Louvain, (1969) (Versión Española: Tecnos, Madrid).
2. B. L. Spencer, *Laws of Form* (Dutton, London, 1969).
3. F. G. Varela, 'A Calculus of Self-Reference', *Int. J. General Systems* **2**, 5–24, (1975).
4. M. Eigen and P. Schuster. *The Hypercycle: A Principle of Natural Self-Organisation* (Springer, Heidelberg, 1979).
5. C. O. François, 'Théorie Générale des Systèmes Disperses' 9è Congres Int. de Cybernétique, Namur (1980).
6. H. Odum, *Environment, Power and Society* (Wiley, New York, 1971).
7. G. Nicolis and I. Prigogine, *Self-Organization in Nonequilibrium Systems* (Wiley, New York, 1977).
8. M. Eigen and P. Schuster, *The Hypercycle, Op cit.*
9. J. P. van Gigch, *Applied General Systems Theory* 2nd Ed. (Harper & Row, New York, 1978).
10. C. O. François 'Un Enfoque Cibernético de las Sociedades Humanas' *Rev. Inst. Cib* **3**, 1–2 (Buenos Aires, Argentina, 1978).
11. R. Boudon, *La Place du Désordre* (Presses Universitaires de France, Paris, 1984).
12. Y. Barel, *La Reproduction Sociale* (Anthropos, Paris, 1973).

SECTION III

EPISTEMOLOGIES

6. COMMUNICATION BETWEEN MINDSCAPE TYPES*

Magoroh Maruyama

The author conveys the wisdom of engaging in transepistemological understanding which consists of avoiding the entrapment of our thinking within categories of theories, logics, epistemologies and/or disciplines. The characteristics of four epistemologies are illustrated.

EPISTEMOLOGIES AND MINDSCAPE TYPES

In my previous publications I have discussed 'mindscapes' and 'Mindscape types'.[1–5] Mindscapes can be grouped into different types, such as H, I, S and G. Mindscapes within the same type share the same structure, even though their topical focus and contents may be different. Communication and understanding between mindscapes within the same type are quite straightforward, while those between different types are difficult and often impossible. In this chapter, I would like to address myself to the problems in the latter.

There has been confusion, among the readers of my previous publications, between 'mindscapes' and 'mindscape types'. In order to minimize this confusion, I would like to use in this chapter the term 'epistemologies' in lieu of 'mindscape types'.

THEORIES, LOGICS AND EPISTEMOLOGIES

Theories, logics and epistemologies constitute three distinct levels. Different theories can be built on the same logic, different logics may stem from the same epistemology, and there are different epistemologies (see Figure 1). I shall not define these levels, because the concept of definition is valid in one epistemology but not necessarily in others. Rather, I shall give examples. Euclidean geometry and non-Euclidean geometries are different theories within the same logic, which is deductive and bivalued. This bivalued deductive logic, commonly called 'Aristotelian logic', and multivalued deductive logics, often erroneously called non-Aristotelian logics, are based on the same epistemology. This epistemology is deductive and has a hierarchical universe in which values can be rank-ordered, usually obeying the transitive law. In contrast, the concept of reciprocal causal loops belongs to a non-hierarchical epistemology. Reciprocal causal loops, which are abundant in biological, social, and even some physical processes, differ from non-reciprocal causal

* Update and revision by the author of "Transepistemological Understanding: Wisdom Beyond Theories", In: R. Hinshaw (Ed.) *Sol Tax Festschrift (Currents in Anthropology)*, (Den Haag, NL, Mouton Publishers, 1979.) Used with Permission. Copyright 1979, Mouton Publishers (Division of Walter de Gruyter, Berlin).

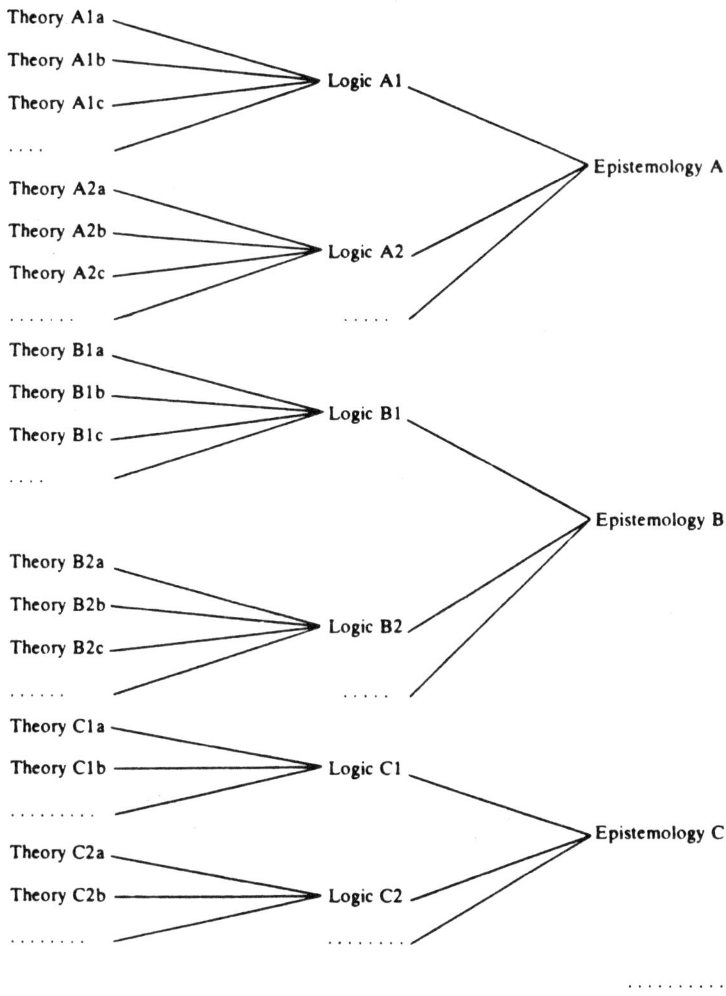

Fig. 1. Theories, Logics and Epistemologies.

processes even mathematically. This explanatory three-level structure may look hierarchical, but it is so presented only for the convenience of those who are used to hierarchical thinking. For those who do not need hierarchical thinking, the structure can be explained differently.

Of the many epistemologies, the following four are useful for illustration: homogenistic, independent-event, homeostatic and morphogenetic. I have summarized the characteristics of these four epistemologies in Table 1, Appendix 1. This is an oversimplification; I have discussed the problem in more detail elsewhere.[6] It is also excessively categorical; the categories should not be taken too categorically, or we will fall into the trap in which the hierarchists now find themselves. They are meant to be neither exhaustive nor mutually exclusive. Two of them, the homogenistic epistemology and the independent-event epistemology, will be fairly familiar to Western readers because they are two sides of the same coin, mistakenly called 'Western logic'. The homeostatic and morphogenetic epistemologies may seem exotic, but they correspond to biological and social processes of equilibrium, increase of differentiation, growth of patterns and structures, heterogenization, symbiotization and evolution.

In a sense, the table serves as an epistemological Rorschach test. Those who adopt the homogenistic epistemology invariably ask whether the categories are universally valid, exhaustive, and mutually exclusive, find fault with the Table in terms of these criteria, and attempt to construct universally valid, exhaustive, and mutually exclusive categories. Those who adopt independent-event epistemology are relatively uninterested in the Table. Those with homeostatic and morphogenetic epistemologies proceed on the assumption that the construction and interpretation of tables depend on the situation, and consequently they do not worry as to whether the categories are universally valid, exhaustive, or mutually exclusive. In fact, they consider these criteria irrelevant and useless. The difference between those who adopt the homeostatic epistemology and those who adopt the morphogenetic epistemology is that the former will seek a stabilised pattern between differing tabulations and interpretations by different persons, while the latter will expect change over time. For example, my epistemology happens to be morphogenetic: in the first place, I need no neat categorization in my head, and in fact it is unnatural for me to categorize; furthermore, if I make categorizations for the convenience of others, they come out different each time. This is a comment on how people with different epistemologies look at the Table differently and is therefore a metacomment, which, incidentally, agrees with the Table.

The contents of the Table are to a great extent self-explanatory, with the exception of reciprocal causality, which requires some additional discussion because of its recent history and ramifications. Within Western thinking, reciprocal causality was scientifically recognized only in the 1940s, when mathematical sophistication developed with the study of various feedback systems such as radar-controlled anti-aircraft artillery and servomechanisms. Until then Western logic had prohibited the formal formulation of causal loops, and non-reciprocal causal models had, with occasional exceptions, dominated Western thinking for almost 2,500 years.[7,8] Since 1940, reciprocal causal models have undergone a series of developments. In its first phase, during the 1940s and 1950s efforts were concentrated on deviation-counteracting, stabilising reciprocal causal processes such as homeostasis, morphostasis, and servo-mechanism. In its second phase, from around 1960 to the present, differentiation-amplifying reciprocal causal processes began to draw scientists' attention.

In the course of the first phase, there was increasing realization that C. Shannon's information theory was inadequate[9]. This theory per se had nothing to do with reciprocal causality. In fact, it was based on the independent-event epistemology, in the sense that it assumed the basic state of the universe to be homogeneity and considered heterogeneous structures as unusual occurrences which tended to decay into homogeneity under the influence of random noise. In Shannon's formulation, nothing could grow. It was gradually recognized that what increases structures, differentiation, and heterogeneity in such processes as evolution, is another type of reciprocal causal process. An example of this kind of process is the case of a particular species of moth and the particular species of bird that feeds on it. Camouflaged mutants of the moth survive better, and the mutants of the bird that are more capable of finding the camouflaged moths survive better. Therefore, the moth gets more and more camouflaged generation after generation, and the bird gets more and more capable of finding the camouflaged moth. This type of reciprocal causal process has been called 'deviation-amplifying', 'morphogenesis', 'positive feedback system', and so on. Maruyama[10] and Ulam[11] mathematically demonstrated that simple rules of interaction can generate complex patterns and that it takes more 'information' to describe the finished patterns than to describe the generatiing rules. In other words, the amount of information, as defined by Shannon's formula, can grow.

Just as reciprocal causal processes can increase the amount of Shannonian information, they can also systematically increase the unevenness of distribution of temperature in the biological and social world. This is impossible under the assumption that the basic state of the universe is random distribution of independent events. To clarify this notion, we must examine two misconceptions, one regarding the relationship between thermodynamic entropy and information and the other regarding the relationship between thermodynamic entropy and energy.

Thermodynamic entropy is a measure of thermal homogeneity and is defined in such a way that it is greater when temperature is more evenly distributed. The fact that it increases with thermal homogeneity, while the amount of Shannonian information decreases with structural homogeneity, is purely definitional. Shannon could have defined the measure of the amount of information to increase with structural homogeneity in order to parallel the definition of thermodynamic entropy. Alternatively, thermodynamic entropy could be redefined to decrease with thermal homogeneity in order to parallel Shannon's definition of amount of information. In any case, there are no grounds for maintaining any mystical notion regarding the fact that the Shannonian measure of information has a polarity opposite to that of thermodynamic entropy. Furthermore, since thermodynamic entropy is a measure of *thermal* homogeneity, while the Shannonian measure of information deals with *structural* homogeneity, they are not directly related; nor are they convertible to or from each other.

The misconception regarding the relationship between entropy and energy is of a different nature. This misconception usually occurs in one or both of two forms: (a) as if entropy and energy were convertible to and from each other and (b) as if they added up to a fixed sum, that is as if what is gained in one were lost in the other.

The first law of thermodynamics is the law of conservation of energy. The second law is the law of increase of entropy, in other words, the law of thermal homogenization. These laws are logically and physically independent. Furthermore, there is no law of conservation of entropy. If entropy increases somewhere, it does not have to decrease somewhere else. In fact, one form of the second law of thermodynamics says that, in an isolated system, entropy has a high probability of increasing. It is important to note that entropy increases within the system without decrease of entropy elsewhere; and there is neither energy input nor energy output during this process of increase of entropy; and that total energy remains constant within the system even though entropy changes.

Entropy and energy can change independently. There is no direct conversion between entropy and energy. Furthermore, the second law of thermodynamics applies only to temporal stochastic, causally non-reciprocal processes, not to reciprocal causal (both deterministic and probabilistic) processes. Therefore it is not logically contradictory for entropy to decrease within an isolated system if there are reciprocal causal processes going on in that system.

As we have seen, the misconception regarding the relationship between entropy and information is mainly due to insufficient understanding of the definitional nature of these terms. On the other hand, the misconception regarding the relationship between entropy and energy is aggravated by an epistemology based on the notion of substance, which tends to treat entropy and information as substances of quantities which persist. Energy is a quantity which persists in various forms and which is even convertible to and from matter. Entropy and information, in contrast, are forms and structure which do not persist, and they are not convertible, from one to the other. Reciprocal causal models so far have succeeded in explaining growth of patterns, heterogeneity, and information. The next problem to be studied is the process of symbiotization once heterogeneity has been generated. We have not worked out methods for producing positive-sum combinations of heterogeneous elements which are not yet interacting. If Individual A has three ways a_1, a_2, a_3, to implement his goal and Individual B has two ways b_1, b_2, to implement his, there are $3 \times 2 = 6$ different alternative combinations between the two individuals. Some of these combinations may be positive-sum, others not. For a greater number of individuals, the number of combinations becomes enormous. There may be individuals who cannot be part of the same network because no symbiotic combinations between them can be found. The selection of positive-sum combinations and the selection of networks for each individual require methods we do not yet have.

Through the development of reciprocal causal models, it has become clear that the basic principle of biological, social, and even some physical processes is increase of heterogeneity and symbiotization. Westerners have traditionally regarded homogeneity as the basis of peace and heterogeneity as a source of conflict. In biological processes, however, it is often homogeneity that causes conflict while heterogeneity confers benefits. For example, animals convert oxygen into carbon dioxide, and plants convert carbon dioxide into oxygen. In doing exactly opposite things, they do not conflict; on the contrary, they help each other. The richness of life in the tropical rain forest or on the coral reef is possible because of the heterogeneity of the species living there. If all animals were to eat the same food, there would be food shortage. If not all animals were eaten, there would be waste. Heterogeneity also assures richer and faster evolution and a higher probability of survival in the case of ecological and environmental change. Resource diversification, risk dispersion and positive-sum relations are three benefits of interaction among heterogeneous elements.

The Mandenka of West Africa say: 'If you force people to be similar, the only way left for them to be different is to get on top of one another. This creates conflicts'.[12]

Traditionally, Westerners have had to choose between the two sides of the same coin (occasionally with various degrees of mixture): the homogenistic epistemology and the independent-event epistemology. Examples of this limited choice are the medieval controversy between universalism and nominalism, with an intermediate position taken by conceptualism, and the 19th and 20th century contrast between idealists such as Kant and Hegel and existentialists such as Kierkegaard, Heidegger, and Sartre. The same limitation is found in many of the conflicts in the social sciences today. In sociology, the quantitative school reflects the homogenistic and the phenomenological school the independent-event epistemology, although symbolic interactionism has moved toward the morphogenetic epistemology. In anthropology, the universalists' arguments belong to the homogenistic and most of relativists' arguments to the independent-event epistemology. (Some of the latter, being quantitative, reflect the homogenistic.)

It is incorrect, however, to identify the homogenistic and independent-event epistemologies as 'Western', because they are predominant in many non-Western cultures as well. Islam is more hierarchical and categorical than Christianity. The Hindu philosophy is also hierarchical and homogenistic. On the other hand, Mandenka epistemology is morphogenetic, and Chinese philosophy is homeostatic. Japanese philosophy has at least three underlying currents; the Jomon current, which originated around 7000 B.C., is morphogenetic; the Yayoi current, which began about 300 B.C. is homeostatic; and the Yamato current, which came later via Korea from horse-riding hunters of northern Asia, is hierarchical. The Jomon current accounts for the Japanese's view that nothing remains the same and hence their readiness to change. (This contrasts with the traditional Chinese philosophy of oscillation, in which there is a return to the same state.) The Yayoi current is responsible for the Japanese's receptiveness, absorption, and perfectionism, and the Yamato current is seen in their hierarchy-consciousness. To add to the heterogeneity within the Japanese culture, the three currents have been distributed differently in different social classes. Up to and including the feudal period, the small ruling class was hierarchical and homeostatic, while the grass-roots culture of farmers was egalitarian and reciprocally interactive. After the Meiji reform, the Yayoi and Yamato epistemologies spread into the growing middle class, although the Jomon current remained in all social classes. Each person in the Japanese culture incorporates all three currents. In contrast, the Western individual chooses to belong to one religion, one ideology, or one political party. The interaction between the Jomon and the Yayoi currents is also expressed in various forms of traditional and modern Japanese architecture.[13,14]

Furthermore, Japanese and Mandenka philosophies are heterogenistic, while Chinese philosophy is dualistic. In this respect Chinese philosophy resembles Greek logic on the surface, and it is this superficial resemblance which has caused Europeans to interpret Chinese philosophy as having a 'logical' structure, even though this interpretation contains some distortions. Similarly, Europeans have recognized the Islamic and Hindu philosophies as 'great philosophical systems' because they are hierarchical and homogenistic.

In summary, the Islamic, Hindu, and Chinese philosophies, which Westerners consider 'typically Eastern', have affinities to Greek philosophy. Jomon, Mandenka, and to some extent Navajo and Invit philosophies form a different cluster and have not been recognized as 'philosophical systems' by Western philosphers.

DIMENSION REDUCTION

Mutual understanding between persons with different epistemologies is often illusory. Quite often a person believes that (s)he 'understands' other persons, while in fact (s)he is only reducing their thinking to her/his own epistemology. I shall call this process *dimension reduction*. The conversion of one epistemology to another may or may not produce an internally consistent 'explanation'. When it does, the person is likely to be convinced that the explanation is the right one even if it is incorrect. I shall illustrate this with two metaphors. For the sake of simplicity, these metaphors are at the level of theories rather than at the level of epistemologies.

The first metaphor is the two-dimensional, parabolic movement of a ball thrown into the air (ignoring the resistance of the air and the rotation of the earth). The situation is illustrated in Figure 2 where the *X*-axis represents the horizontal direction in which the ball is thrown and the *Y*-axis represents the vertical direction. We can make a dimension reduction of this movement in several ways.

One way is to project this movement into the *X*-axis in the manner of the shadow of the ball on the ground when the sun is at the zenith. This dimension reduction produces the interpretation that the movement is a one-dimensional movement of a body at a constant speed. This interpretation does not violate the laws of physics and is internally consistent. Therefore the person who makes it may be convinced that it is the correct one.

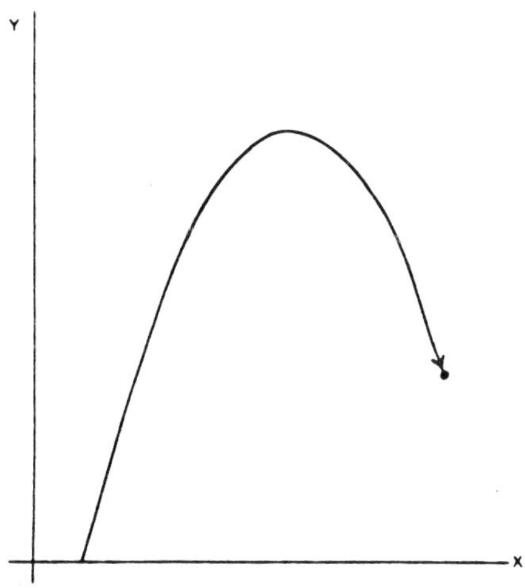

Fig. 2. Movement of a Ball thrown into the Air.

Another way is to project the movement onto the *Y*-axis as it would appear if it were sunset time, the ball was thrown in the direction of the sunset, and we were watching the shadow of the ball on a wall behind the man who threw it. We would see a movement of an object upward with a constant deceleration until it eventually stopped and began accelerating downward. We could interpret this as a one-dimensional movement under the influence of a constant gravitational force. This interpretation does not violate the laws of physics and is internally consistent. Therefore the person who makes it may also be convinced that it is the correct one.

In this instance, the two methods of dimension reduction produce internally consistent descriptions that may induce a delusion of correctness in the interpreter. In the next metaphor, some methods produce internally inconsistent descriptions. Figure 3 represents the movements of two balls on a billiard table. The *X*-axis represents the east-west direction and the *Y*-axis the north-south direction. One ball comes from the east and the other from the west; both are of the same size and weight and have the same speed; they collide in such a way as to bounce off in north-south directions. Again, there are several ways to produce a dimension reduction.

If we project this event onto the line alpha, which runs southwest to northeast, we see two objects move toward each other, collide, and bounce back and move apart, all in one-dimensional movements. This description does not violate the laws of physics and is internally consistent.

If, however, we project this event onto the *X* axis, it appears that two objects fly toward each other, collide, and get stuck, with some decrease in their total size. In this description the kinetic energy gets lost. If we want to preserve the law of conservation of energy, we have to invent some explanation. We can, for example, say that the kinetic energy was converted into potential energy required to hold the two objects together with a reduced size. This is, of course, a rationalization which

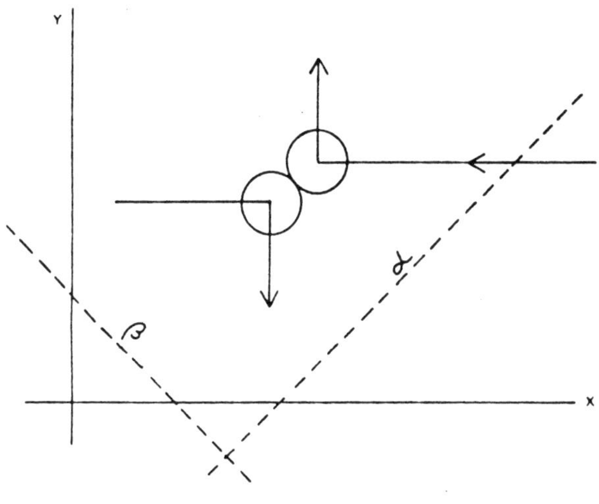

Fig. 3. Movements of Two Balls on a Billiard Table.

produces an internally consistent but incorrect explanation. We cannot exclude the possibility that much of our 'science' consists of such rationalizations.

If we project the event onto the *Y*-axis, one object apparently splits into two pieces that fly away, and the sum of the sizes of the two pieces is greater than the original object. Again, we can rationalize this by saying that an explosion took place and that some potential energy was converted into kinetic energy.

If we project it onto line beta, which runs northwest to southeast, we have a more difficult problem: two objects collide, pass through each other, and fly away. Here we may finally realize that this event is very difficult to explain as one-dimensional, and the second dimension must be added.

Similarly, when a person reduces another person's epistemology to the dimensions available in her/his own, the resulting interpretation may or may not be internally consistent. When the interpretation is internally consistent, (s)he most likely deludes her/himself that (s)he has understood the other person. 'I understand you perfectly, but you keep saying that I don't understand you' is a typical statement made in such situations. When the dimension reduction produces an internally inconsistent interpretation, the person doing the dimension reduction tends to think that the fault is in the other person and to consider the latter illogical, unintelligent, insincere, or deceptive.

SELF-CONSISTENCY, AUTODOX, AND EXOPARADOX

The theory that there is only one logic, or that there is a superlogic to encompass all other logics, is tautological rather than logical. Like solipsism, such a theory is internally self-consistent and therefore self-proving, but not necessarily correct in the sense of correspondence with what it tries to describe as 'reality'. Mathematical logicians know that the self-consistency of a theory is no proof of its universality. In fact, mathematical logicians can produce many self-consistent theories which disagree with one another, for example, Euclidean geometry and some of the non-Euclidean geometries. More important, mathematical logicians know that there is nothing 'objective' about mathematics or mathematical logics; these do not need to correspond to any 'reality'. Mathematical theories and mathematical logics are mental constructs. If there is anything 'objective' about them, it is not in their contents, but in the fact that once you agree to adopt certain axioms and certain rules of deduction, you can arrive at the same conclusions (called theorems).

Social Sciences, on the other hand, aim at producing theories which correspond to 'reality'. Too frequently, however, the internal consistency of a theory leads to a delusion as to its correctness in the theories. The self-consistency of a theory may lead to a tendency unintentionally to distort and reduce other theories to that theory in such a way as to reinforce it. This is exactly the opposite of Russell's paradoxes, which contain self-contradictions*. The tendency I am talking

* An example of Russell's paradoxes is the statement "Everything written on this page is false". If the statement is true, it must be false; if the statement is false, it must be true.

about converts any argument into proof of its own theory. For lack of a better term, I shall call this tendency *autodox*. A self-consistent theory may also provide its opponents with the argument that autodox is predictable in terms of their opposing theories; that is, it may provide opponents with proof that they are right. I shall call this process *exoparadox*. A few examples will illustrate these processes.

A universalist I met at an international congress maintained that all humans had the same basic structure of reasoning regardless of culture. He asked me how I would say 'I see a tree' in Japanese. I made a Japanese translation of that sentence. Then he asked me what that Japanese sentence meant. I translated it back into English and said 'I see a tree'. Thereupon he said that his theory was proved. Upon close examination, many theories are based on this type of rationale. For example, a sociologist who believes in a hierarchy theory of social organization goes to a non-hierarchical culture, constructs a 'measure' or 'scale' of leadership (such as the frequency with which one person talks with other persons), collects 'data', and writes a monograph on the hierarchical structure of the culture. Similarly, the use of the Q-sort technique, in which a person is made to rank-order a number of adjectives, assumes that adjectives, or for that matter everything, can be rank-ordered, or that all persons in all cultures live in a rank-orderable universe. A less technical but more commonly used method is 'value' analysis, in which persons in different cultures are asked to rate certain words, sentences, or situations. From the ratings, the researcher deduces the rank-order of 'values'. The assumption is again that all persons in all cultures live in a universe consisting of ratable events and rankable values, or that all persons in all cultures have a hierarchical epistemology.

The fallacy common to all these examples is that the theorist or the researcher is seeking data from the point of view of her/his own theory, logic of epistemology, forcing her/his epistemological structure upon data and filtering and distorting the data without realizing that (s)he is doing so. (S)he is free of distortion on the level of measurement, but not on the level of epistemology. Autodox is the result of such epistemological distortion. The theorist examines the counter-arguments, re-interprets and epistemologically distorts them into her/his own epistemology, logic of theory, and finds confirmation in them. All (s)he accomplishes is to convince her/himself that (s)he is right. At the same time, (s)he provides others proof that *they* are right and (s)he is wrong. This is what I have called *exoparadox*. The universalist is convinced that everybody has the same thought structure and therefore her/his thought structure is universal. This provides non-universalists proof that the universalist is limited in her/his ethnocentric (non-universal) thinking and therefore even the universalist is non-universal.

Autodox and exoparadox create a trap which confines the theorist in a delusion of the correctness of her/his theory and the incorrectness of others' theories. There is another set of considerations at the epistemological level which contributes to the inability of theories to capture much of the wisdom which exists in different cultures and disciplines: exotication, the unsuspected hole, and self-confinement.

EXOTICATION, THE UNSUSPECTED HOLE, AND SELF-CONFINEMENT

What is self-evident in one epistemology may be incomprehensible in another. For example, in Mandenka wisdom, it is self-evident that heterogeneity and change are sources of harmony and symbiosis while homogeneity is a source of conflict. This is incomprehensible in the homogenistic epistemology, which advocates unity by similarity and considers differences as sources of conflict. Even though such wisdom may be applicable and useful in any culture, those whose epistemology cannot accommodate it may fail to see its applicability and usefulness and consider it exotic, something belonging to somebody else. This tendency may be called *exotication*.

Another example is the concept of reciprocal causality. Even though the concept was understood and used in certain disciplines, such as biology, ecology, and electrical engineering, in many others it remained until recently something exotic and even illogical. Many whose epistemology could not accommodate reciprocal causality attempted to 'understand' it by breaking it down into a series of non-reciprocal causal chains in alternating directions, without realizing that this procedure alters the mathematical characteristics of the description and therefore results in incorrect analysis. Furthermore, those who consider reciprocal causality exotic, regarded it as something that applied only to special cases and failed to see its more general applicability and its full significance.

Still another example involves culturally conditioned sex differences in thought patterns in the United States. For a number of years I taught courses on misunderstandings. I required the students to present case studies of misunderstandings with analysis at different logical levels of communication. Women produced complex cases with sophisticated analysis. When I praised them, the women thought that they did not deserve it because these cases were common everyday occurrences. On the other hand, the men thought that these cases must be fictitious and complained that the women were getting good grades by inventing stories.[15] Here the women's wisdom was exoticated by the men.

Persons in a discipline or a culture many be unaware that others lack the concepts they consider basic. This problem may be called the *unsuspected hole*. For example, ecologists, for whom the concepts of heterogeneity and positive-sum relations are basic, may be unaware that many political scientists and even some sociologists lack these concepts. Similarly, people in cultures with heterogenistic epistemologies may be unaware that Westerners tend to be unable to think of peace in terms other than 'unity by similarity'. As long as a person remains unaware that others lack a concept basic to her/him, it does not occur to her/him to explain it to them, and the concept does not get formulated as a theory.

Persons in a given field may be unaware that a conceptual or epistemological innovation in their field is related to (a part of, resulting from, or affecting) an overall conceptual or epistemological change which includes other fields. This is a failure to see the significance of the innovation in the larger context of intellectual evolution. It is provincialism of a sort, and I term it *self-confinement.*

Sometimes the persons involved are unaware that the innovation is epistemological rather than theoretical or technical, especially if they have been brought up in one epistemology and are unaware of others. In such cases they may be unable to explore the full potential of the innovation and may be trapped in their old epistemology within the innovation. For example, the invention of positive feedback loops in radio engineering occurred about 1910.[16] Even though positive feedback loops are abundant in biological processes and are the basis of differentiation, growth of structure, and evolution, at the time of its invention, the positive feedback circuit was regarded as a local innovation within the field of radio engineering. Furthermore, they were considered a technical rather than a epistemological innovation. They were exoticated by those in other fields.

It is almost ironic that the concept of reciprocal causality gained its visibility through the application of negative feedback loops in the 1940s[17], while positive feedback loops were not given much attention until the 1960s. Perhaps, on the other hand, this was because the concept of negative feedback was more compatible with the epistemology of Western science than that of positive feedback. Reciprocal causal processes of both types were alien to Western thinking. Negative feedback systems, however, with their equilibrium-maintaining characteristics, were much more compatible with Western epistemology, based on the concepts of substance and permanence, than positive feedback systems, whose basic characteristic is change. Therefore, it may be understandable that negative feedback loops found wide recognition in the 1950s while positive feedback loops continued to be exoticated. For example Waddington's (1968–72) pioneering work in theoretical biology still leaned more toward morphostasis than toward the rise of new patterns.[18]

Immediately after the publication of my article 'The Second Cybernetics: Deviation-amplifying Mutual Causal Processes',[19] I received some 70 reprint requests, mostly from geneticists, embryologists, and cancer-research pathologists in Europe and North America. This indicated the disciplinary localization of the concept and self-confinement. In 1968 the article was reprinted in a book for behavioural scientists.[20] Another article, 'Paradigmatology and its Application',[21] drew reprint requests from twelve fields of biological science, eight fields of medical science, four fields of psychological science, three fields of physical sciences, and two fields of social science, indicating a further decrease of localization. In 1976, E. Morin[22] began publishing his series *La Méthode* which ranges from physical and biological sciences to social sciences.

Thus self-confinement is gradually breaking down, with increasing numbers of people realizing that, what they had thought to be a local phenomenon, has wider implications.

In 1963, I wrote to the Editor of *American Scientist* that the concept of deviation-amplifying reciprocal causality would be taken up by social scientists in about five years and by philosophers and general scientists in ten to fifteen years. My 'prediction' turned out to be true. I was able to foresee this because I was familiar with the climate of opinion in various discplines and therefore could estimate the time it would take them to change. The concept of reciprocal causality is the first major conceptual innovation in 'Western logic' in 2,500 years. It will be increasingly appreciated during the coming decades as the cosmological, biological, and anthropological implications of differentiation-amplifying reciprocal causal processes are studied further.

WISDOM BEYOND THEORIES

Theories are limited in several ways. They are limited because they must take a certain logical form in a certain epistemology and because they establish themselves by dimension reduction, self-consistency, and autodox. They are artifical. Much of wisdom cannot be captured by theories.

There are two types of wisdom. The first is the wisdom which exists in specific cultures or in specific disciplines, for example, the Mandenka wisdom. This type of wisdom is often taken for granted within the culture of discipline, and little or no explanation or theorization is needed for insiders. Insiders consider it trivial, and outsiders either are unaware of it or, if aware, consider it exotic, irrelevant and inapplicable to themselves. Therefore, this type of wisdom is reduced to trivialities if attempts at theorization are made, either by insiders or by outsiders. In this sense it is beyond theories. It gives birth to non-trivial theories only under certain circumstances. Mandenka wisdom probably will become theoretically non-trivial because the time is ripe for it: the western world is looking for an alternative logic, and Mandenka wisdom can meet the need. Fortunately, Mandenka wisdom has been written down articulately by Sory Camara because he, as a Mandenka, realized that his French colleagues did not understand it. My articles[23–29] were written under similar circumstances. The wisdom contained in them has existed for centuries in several cultures and in several scientific disiplines. It is trivial in these cultures and disciplines. The need for its articulation arose when I realized that most Westerners could not easily understand what was trivial to me. The time was somewhat ripe in some disciplines, though it will be riper in a wider range of disciplines in a decade or two.

The other type of wisdom is gained in transepistemological understanding. It concerns the limits of applicability of theories, logics, epistemologies, and culture-specific or discipline-specific wisdom, It is a metawisdom, so to speak. The concept of reciprocal causality is wisdom of the first type; my 1963 'prediction' about the rate of acceptance of this specific wisdom, is wisdom of the second type.

We have seen that both types of wisdom are beyond theories. What we need in our world, now and in the future, is wisdom rather than theories. We must improve our transepistemological understanding and avoid entrapment in theories, autodox, and exoparadox.

REFERENCES

1. M. Maruyama, 'Mindscapes and Science Theories' *Current Anthropology* **21**, 589–599 (1980). Reprinted in *General Systems*, XXVI (1981), 41–60.

2. M. Maruyama, 'Denkmuster: metaprinzipien der Umweltgestaltung' *Garten und Landschaft* Oktober, 806–815 (1981).
3. M. Maruyama, 'New Mindscapes for Future Business Policy and Management' *Technological Forecasting and Social Change* **21**, 53–76 (1982).
4. M. Maruyama, "Mindscapes: How to Understand Specific Situations in Multicultural Management", *Asia Pacific Journal of Management* **2**, 125–149 (1985).
5. M. Maruyama, 'Mindscapes, Management and Business Policy' *Academy of Management Review* **7**, 612–619 (1982).
6. M. Maruyama, 'Mindscapes and Science Theories' *Op. Cit.*
7. M. Maruyama, 'The Second Cybernetics: Deviation-Amplifying Mutual Causal Processes' *Amer. Scientist* **51**, 164–179, 250–256 (1963).
8. M. Maruyama, 'Paradigmatology and Its Applications' *Cybernetica* **17**, 136–156, 237–281 (1974).
9. C. Shannon and W. Weaver, *The Mathematical Theory of Communication* (Univ. of Illinois Press, Urbana, Ill. 1949).
10. M. Maruyama 'The Second Cybernetics . . .' *Op. Cit.*
11. S. Ulam, Lecture given at Stanford University (1960).
12. S. Camara, 'The Concept of Heterogeneity and Change among the Mandenka' *Technological Forecasting and Social Change* **7**, 273–284 (1975).
13. K. Tange and N. Kawazoe, *Ise* (MIT Press, Cambridge, Mass. 1965).
14. K. Tange, *Katsura* (Yale Univ. Press, New Haven, Conn. 1972).
15. M. Maruyama, 'Basic Elements in Misunderstandings' *Dialectica* **17**, 78–92, 99–110 (1963).
16. J. Milsum (Ed.) *Positive Feedback* (Pergamon, Oxford 1968).
17. M. Maruyama, 'The Second Cybernetics . . .' *Op. Cit.*
18. C. H. Waddington (Ed.) *Towards a Theoretical Biology*, Four Volumes (Aldine, Chicago, 1968–1972).
19. M. Maruyama 'The Second Cybernetics . . .' *Op. Cit.*
20. W. Buckley *Modern Systems Research for the Behavioral Scientist* (Aldine, Chicago 1968).
21. M. Maruyama 'Paradigmatology and Its Applications' *Op. Cit.*
22. E. Morin, *La Méthode* (Seuil, Paris 1976).
23. M. Maruyama 'Morphogenesis and Morphostasis' *Op. Cit.*
24. M. Maruyama 'The Second Cybernetics . . .' *Op. Cit.*
25. M. Maruyama 'Paradigmatology and Its Applications' *Op. Cit.*
26. M. Maruyama 'Heterogenistics' *Acta Biotheoretica* **26**, 120–136 (1977).
27. M. Maruyama 'Heterogenistics and Morphogenetics' *Theory and Society* **5**, 75–96 (1978).
28. M. Maruyama 'Mindscapes and Science Theories' *Op. Cit.*
29. M. Maruyama 'Denkmuster : metaprinzipien . . .' *Op. Cit.*

APPENDIX 1

Table 1. Characteristics of Four Epistemologies

	Type H Homogenistic Hierarchical Classificational	Type I Heterogenistic Isolationistic Random	Type S Heterogenistic Interactive Homeostatic	Type G Heterogenistic Interactive Morphogenetic
Philosophy	*Universalistic:* Abstraction has higher reality than concrete things. *Organismic:* Parts should be subordinated to the whole.	*Nominalism:* Only individual elements are real. Society is merely an aggregate of individuals.	*Equilibrium or cycle:* Elements interact in such a way as to maintain a heterogeneous pattern or go in cyclic change of such pattern.	*Heterogenization, symbiotization, and evolution:* Heterogeneity is indispensable and basic to biological and social processes.
Causality	Two things cannot cause each other. Cause-effect relations may be deterministic or probabilistic.	Independent events are most natural, each having its own probability. Non-random patterns and structures are improbable and tend to decay.	Reciprocal: Deviation-counteracting (negative feedback loops), both probabilistic and deterministic.	Reciprocal: Differentiation-amplifying (positive feedback loops), both probabilistic and deterministic.
Logic	Deductive, axiomatic. Mutually exclusive categories. Permanence of substance and identity.	Inductive, statistical.	Complementary.	Heterogeneity, harmony, change, unrepeatable and irreversible processes.
Perception	Rank-ordering, classifying, and categorizing into a neat scheme. Seek regularity.	Isolating. Each is unique and unrelated to others.	Contextual. Seek balance, stability.	Contextual. Seek new interactions and new patterns. Things change and relations change; therefore meanings change and new meanings arise.
Knowledge	Belief in existence of one truth. If people are informed, they will agree.	Why learn beyond my own interest?	Polycular: Binocular vision enables us to see three-dimensionally, because the *differential* between two images enables the brain to compute the invisible dimension. Cross-subjective analysis enables us to compute invisible dimensions.	
Information	Past and future inferable from present probabilistically or deterministically. The more specified, the more information.	Information decays and gets lost. Blueprint must contain more information than finished product.	Loss of information can be counteracted by means of redundancy.	Complex patterns can be generated by means of simple rules of interaction. Amount of information can increase.

Table 1 (*continued*)

	Type H	Type I	Type S	Type G
Cosmology	Causal chains. Hierarchy of categories, subcategories, and supercategories.	The most probable state is random distribution of events with independent probability. Conditional probability is less usual.	Equilibrium by means of mutual corrections, or cycles due to mutual balancing.	Generating of new patterns through interaction.
Processes	Repeatable if conditions are the same.	Structures decay. Universe tends to homogeneity.	Structures maintain.	Structures grow. Heterogeneity, differentiation, and symbiotization increases.
Aesthetic principles	Unity by similarity and repetition. Dominant theme is reflected in subdominant themes.	Random, capricious, haphazard.	Balance, completed equilibrium. Static harmony of diverse elements.	Changing harmony of diverse elements. Avoidance of repetition. Design toward multiple interpretations. Deliberate incompleteness to enable addition of new elements. Dynamic harmony.
Ethics	Competition. If not same, then conflict. Let the "strongest" dominate homogenistically. Majority rule (domination by quantity).	Isolationism. Self-sufficiency. If poor, own fault.	Avoid disturbance. One change must be counteracted by another. Symbiosis.	Symbiotization: Look for positive relations with aliens. Regard differences as beneficial. Change: Incorporate new endogenous and exogenous elements.
Religion	Monotheism: Creator and prime mover, omnipotent, omniscient perfect god. Missionary work to convert others (belief in superiority of own religion).	Individual beliefs.	Static polytheism: Maintain established harmony of diverse elements.	Dynamic polytheism: Harmony of diverse elements. Look forward to new harmony when new elements are added or new events occur.

Table 1 (*continued*)

	Type H	Type I	Type S	Type G
Beauty	Unity by repetition, similarity and symmetry. Geometric regularity is beautiful. *Examples:* Gothic, Islamic architecture. French, Italian gardens. Chinese and Japanese Yamato architecture.	Each building or element is unrelated to others. Random, capricious, haphazard. *Examples:* Manhattan, Los Angeles downtown.	Avoid repetition of similar elements. Static harmony of diverse elements. Asymmetry. Completed equilibrium which cannot be disturbed. *Examples:* Some of Japanese gardens and floral art. Japanese Yayoi culture.	Changing harmony of diverse elements. Avoid repetition and symmetry. Designed for simultaneously multiple as well as changing interpretations. Deliberate incompleteness to enable addition of new elements and changes. *Examples:* Some of Japanese gardens and floral art. Japanese Jomon culture.
Space	Defined as between masses, between points, along shafts. Space is a transparent mass, has shape, volume, boundary, identity, specialized function. Space opposes mass and other spaces.	Unrelated units.	Miniature universe containing all phenomena. Internal interaction of heterogeneous elements. Space represents interrelations and harmony, which may be steady or cyclic.	Locality with its own quality and feelings. Connected to other localities. Often thought of as around something (rock, grove) which is considered to be condensation of the quality which permeates the locality.
Objects	Volume and mass opposing space, and opposing one another. Often mass represents human; space represents nature.	Each is its own expression.	Represent phenomena and elements of the universe. Objects support one another, and do not oppose space. Object is quality, not quantity or mass.	Represent growing process, vitality, vicissitude. Objects stimulate one another. Interplay of space and objects. Object is agility, not substance. Space is movement.
Time	Buildings are designed to persist in time. Permanence. Buildings embody eternal principles.	Permanence is unimportant.	The form persists while the building materials may be renewed from time to time. Include seasonally changing elements to emphasize each season and annual cycle.	Increasing heterogeneity and symbiotization. Design permits addition of new elements, pattern change. *Example:* Katsura Palace.

Table 1 (*continued*)

	Type H	Type I	Type S	Type G
Conceptual movement between objects	Tension and extension. Straight lines spanned between points. Lines and surfaces of mass extended to create shafts of space. Lines radiating from or converging to a point. Curves created by physical contour of mass.	Caprice. Random.	Stability. Triangles of unequal sides (not physically present but conceptually underlie between elements). *Example:* Some of Japanese floral art.	Flow. Spiral or curve suggested by relative position of objects, but not by objects themselves. Curves extend and do not return. *Example:* Some of Japanese floral art.
Perceiver's movement	Pre-planned and sequential. Perceiver maintains his identity as a point which moves in space.	Random.	Perceiver permeates simultaneously in all parts, and feels invisible parts and behind objects. The entire design is perceived simultaneously, not sequentially.	Generated by perceptual experience which has multi-branching effect, not single-tracked.
House	Separate the inside from the outside.	Separate one household from another.	Continuation of outside into inside. Removable outer shells. Garden continues into house. Floor extends to outdoors. River flows under floor. Passive appreciation of nature and environment. Absorb the outside into the inside.	A base for activity and interaction with environment, such as a Japanese farm house.
Room	Specialized rooms (bed room, dining room etc.) occupied by specialized furniture. Rooms are separated by walls and can be locked.	Specialized to individual user's needs rather than by generally categorized functions.	Continuity, connectability, convertibility and self-heterogenization. Partitions between rooms can be removed. The same room can be used for different functions. Furniture is stored away and brought out for varying functions. *Example:* traditional Japanese houses.	Continuity, connectability, convertibility and self-heterogenization. Partitions between rooms can be removed. The same room can be used for different functions. Furniture is stored away and brought out for varying functions.

Table 1 (*continued*)

	Type H	Type I	Type S	Type G
Privacy	Superior can monitor subordinates, but subordinates may not see superiors' intention or activities. Concern with patent, copyright etc. based on zero-sum game assumption.	Visual, acoustic and other insulation for each individual. Insulation is considered to increase efficiency. Negative-sum game assumption.	Maximum sharing of intimate concerns. Group solidarity. Positive-sum game assumption.	Less concern with privacy, patent and copyright, because interaction is seen as positive-sum game which generates new inventions.
Sociability	Socialize within homogeneous group. Hierarchy of groups, subgroups and supergroups.	Self-containment. Freedom from social obligations and freedom for individual caprice.	Mutual dependency. Concern over not disturbing the existing harmony. Perpetuation of familiar relations.	Make new contacts, new network. Explore new patterns. Go beyond own group.
Meaning	Neat categories, clear without context.	Individual meaning.	Meaning is contextual and interdependent.	Ambiguity is basic to further development and change.
Design choice	There is "the best design" for all persons.	Give each person a cubicle where he/she can make own design.	Different designs preferred by different individuals can be aesthetically combined in an interesting whole.	Individuals stimulate one another by their varying designs. Experiment with new combinations.
Decision process	Majority rule, consensus or "educating" to make people appreciate the "best" thing.	Do your own thing. Let others alone.	Compensation for hardship on any single individual regardless of which design is chosen.	Compensation for hardship on any single individual regardless of which design is chosen.

7. SYSTEMS, MODELS, METHODS, EPISTEMIC INSTRUMENTATION

B. H. Voorhees

This paper presents a model of the scientific enterprise in the context of a general model of human action. In the view presented, the action of science involves three independent epistemic paths called rationalism, metaphorism, and empiricism interacting in an ongoing process. The terms system, model and method are defined and their role in this process indicated. The result is an extension of the well known hypothetico-deductive process to include the explicit recognition of the importance of scientific paradigms.

INTRODUCTION

The intellectual and conceptual horizons of western science have been greatly extended in the past two decades by the recognition that at least some of the traditions of antiquity, in particular those of magick[1,2] and mysticism,[3] are not only rationally comprehensible but bear many points of correspondence with our own scientific tradition. This recognition has received concomitant and increased scientific interest with such issues as consciousness,[4,5] the mind/body problem[6] and clinical research carried out with various psychoactive drugs.[7] At present, however, there is no clear recognition of the way in which the magical, mystical and scientific currents interact; either within individuals or in society. Rather there is much mis-information, resulting in unwarranted prejudices, blocking legitimate efforts at understanding.

The best remedy for this is, of course, accurate information. The word 'information,' in spatio-visual analogy, calls forth the image of an orderly arrangement of form. The measure of information depends on the improbability of the observed order, relative to other possible orderings of the same forms. In verbally transmitted information the order is temporal rather than spatio-visual. With charismatically transmitted information the order is spatio-temporal. Learning to read and write are the discipline of building appropriate linkages between these and other forms of information processing.

Anticipating a later definition, the purpose of this paper is to introduce a *model* of the scientific subsystems of the system of human action. Here, *model* is understood as a rational construct used as a tool in analysis. The criteria of validity, however, is neither rational nor empirical. It is practical: are informed investigators who make use of the model able to derive some degree of understanding within their own rational-empirical universe?

In terms of application one attempts to fit known data to the model, accepting the model as-if-true. If this proves fruitful: if large chunks of data suddenly fall into place; if new and potent insights result, then the model itself gains credibility. It will be used by more investigators. At some point it will come to be accepted as a valid metaphor by a scientific community, although not all members of that community will necessarily use it — there are individual differences in preference.

The model described in this paper is a part of a larger model of the system of human action synthesizing the triple of science, magick and mysticism. Both of these models are self-referential in that they contain, and give meaning to, the word model. Since self-reference is a characteristic of human consciousness, it is unavoidable in any model of human thought or action which claims completeness. In order to place the model of science presented in semantic context, I begin with a brief discussion of the larger model of action. This discussion defines science functionally (in terms of its contribution to action), and relationally, in terms of its relation to the subsystems of magick and mysticism. A complete description of the model of action will be published elsewhere.

THE SYSTEM OF ACTION

The basic hypothesis is that all human action may be studied in terms of three mutually interdependent components: a knowledge base; certain available means; and an orientation. With this general philosophical presupposition the science subsystem generates a knowledge base, the magick subsystem provides means for action, and the mystical subsystem provides orientation. The system itself is modelled by the diagram of Figure 1.

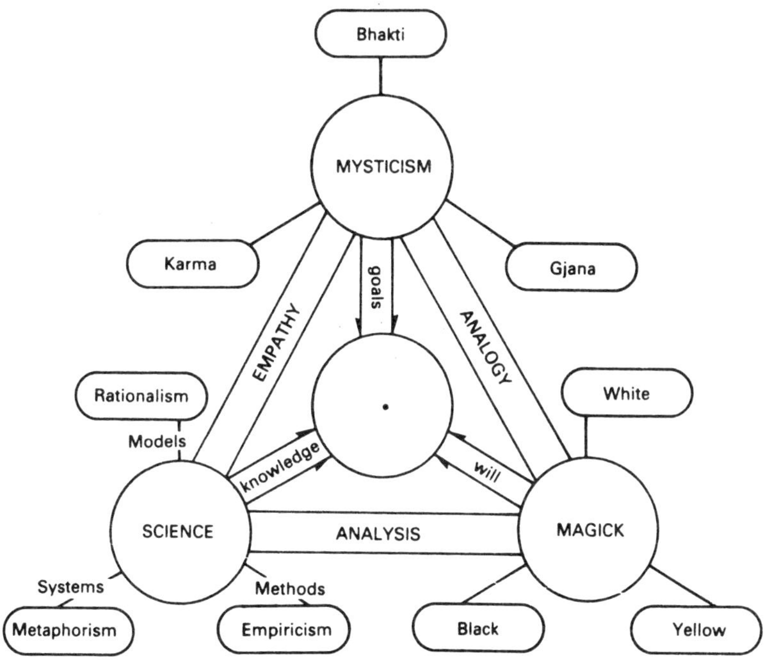

Figure 1.
A Model of Action

The term science is used in the sense of Goldstein and Goldstein.[8] That is, *science is a search for understanding, expressed in terms of general, experimentally tested, laws and principles.* The terms of magick and mysticism are technically defined as well. *Magick is 'the art of producing change in conformity with the will'* and *mysticism is the pursuit of human self-realization.*[9]

The arrows connecting each of these subsystems to the central circle are labeled with the primary contribution which that subsystem makes to the field of action. Since the nature of this action is not specified, the circle is left empty except for the central point. This symbolizes the human agent within an environment. The mystical, scientific and magical subsystems are connected by paths labeled with the particular form of reasoning shared by linked subsystems. Thus, empathy is employed in both science and mysticism, analysis in science and magick, and analogy in magick and mysticism. In each case one form of reasoning is not directly linked to a subsystem, and is treated differently within that subsystem than the other two. Sciences use empathy and analysis to study the validity of analogies with the goal of transcending them by discovery of their limitations. The particle paradigm of light, for example, provided an analogy for thinking about light in terms of physical particles. Mysticism uses empathy and analogy in its acknowledged goal of transcending analysis to achieve a state of 'reunion'. Magick uses analysis and analogy to produce change in a process which involves the will-directed making and breaking of empathetic bonds.

Finally, the actual operation of each subsystem is seen as involving an on-going process of interaction between the three indicated sub-systems. This process will be explicated for the science subsystem in the remainder of this paper. Particular attention will be given to the function of systems, models, and methods in this process, as indicated in my title.

THE SCIENTIFIC SUBSYSTEM

The model presented for the science subsystem involves the on-going interaction of the three sub-systems of rationalism, metaphorism and empiricism in a process called the psycho-epistemic process.[10] This is posited as giving a more complete picture of the scientific enterprise than the hypothetico-deductive process of the logical positivists. The works of Feyerabend[11] and Kuhn,[12,13] have demonstrated the incompleteness of this latter model, based, as it is, on a rational-empiricist epistemology.

The hypothetico-deductive process is illustrated in Figure 2. According to this figure hypotheses are generated by induction from empirical data. Predictions based on these hypotheses are obtained deductively and tested against further data. This dialectic was

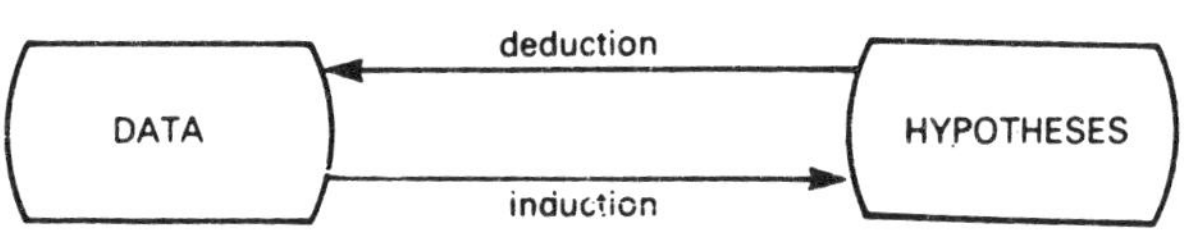

Figure 2.
The Hypothetico-deductive Method

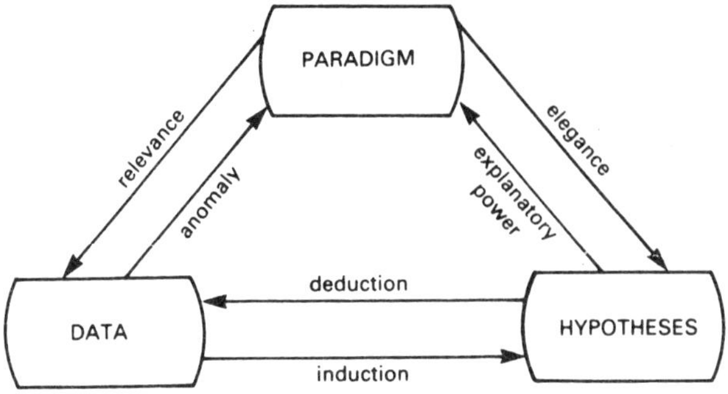

Figure 3.
The Psycho-epistemic Method

assumed to converge to a hypothesis which, although it could not be taken as true, could not be falsified[14] within the limits of existing empirical technique.

With the fall of logical positivism, a controversy has occurred as to what is an appropriate philosophical position for an epistemology of science. Feyerabend proposed anarchy as does Weimer,[15] who claims that the ultimate arbiter of truth is rhetoric. Royce[16] suggests theory laden constructivism. Tolman[17] attacks all forms of constructivism and argues in favour of dialectical materialism — unaware that this leads back to some variety of logical positivism.

The model of science given in this paper is based on Royce's psychological epistemology. Royce distinguishes three independent ways of knowing which he calls rationalism, empiricism and metaphorism. The respective criterion of validity for these three epistemic paths are logical consistency, replicability, and universality. Thus a metaphoric truth claim is based on the consensus of an informed community of knowers. According to Kuhn this is precisely the basis on which a scientific paradigm is accepted or rejected. Thus the introduction of metaphorism augments the rational-empirical epistemology to include Kuhn's ideas of paradigm science and revolutionary science. Based on this, Voorhees and Royce[18] have suggested that the hypothetico-deductive method as indicated in Figure 2 be changed to the psycho-epistemic method shown in Figure 3.

This indicates that induction of hypotheses and deduction of predictions only occur within the context of a given paradigm. The paradigm determines which data is relevant and why hypotheses are elegant. Accumulated data acts to change a paradigm through the appearance of recurrent anomalies, while hypotheses act to stabilize a paradigm by providing explanatory power and attempting to reconcile paradigm and anomaly.

SYSTEMS, MODELS, METHODS

A system is taken to mean a whole which is simultaneously conceptualized as being composed of subsystems, sub-subsystems and so on, together with their interactions. Thus it is simultaneously one and many. Following Zeleny, it is a unity of its structure and organization. This places it beyond both formal logic (because a unity, logically, cannot be composed) and dialectics (because it is a functional unity, which is a dialectically incomprehensible concept). Systems, as defined above, can be conceptualized in terms of Ichazo's[19,20] trialectical logic. Elsewhere[21] an axiomatic mathematical approach to such systems has been presented. The primacy association in Figure 1 of systems with a metaphoric epistemology is that the system must be perceived as a whole before we can agree that it is, indeed, a system. That is, *a system must satisfy the criteria of universality before it can be studied as a system.*

A model is defined as a rationally constructed metaphor of a system or of one of its components, used as a tool for analysis.

The primary association is with rationalism. We can agree, for example, that there is such a thing as the world system, but in order to study this system we must proceed by constructing models to serve as tools of rational analysis.

A method is defined as a rational algorithm which prescribes a set of procedures whose implementation constitutes a process resulting in a specific empirical result. Both the hypothetico-deductive and psycho-epistemic methods, for example, claim the generation of scientific knowledge as result. The primary association of method with empiricism occurs because methods, at least in science, are based on generation and/or analysis of empirical data.

Notice that models are metaphors used as tools of rational analysis; methods are rational procedures used for empirical analysis; and systems are empirical objects or things in the world defined by common agreement for metaphoric purposes. Thus models provide a bridge between metaphorism and rationalism; methods, a bridge between rationalism and empiricism; and systems, a bridge between empiricism and metaphorism.

With these definitions it becomes possible to give a clearer description of the way in which the psycho-epistemic process of Figure 3 generates scientific knowledge. The process begins neither with data, as the empiricists and logical positivists would have it; nor with theory, as posited by theory-laden constructivists. Rather, it begins with recognition that a particular system exists, and that certain conceptual models may be useful in its study. This begins as metaphor and becomes paradigm with the choice of a unique set of models for the system and its components. Prior to this choice no effective analysis is possible. This initial state corresponds to what Kuhn[13] calls the pre-paradigmatic stage of a science. Using the example of physical optics he argues that in the pre-paradigmatic stage a science does not advance. Ichazo[19] makes a very clear statement of this: 'A culture is born only when the unity has been discovered. . . . Before this there are separate groups in a tremendous territorial fight and nothing more.'

The recognition of a system includes the recognition that the system may be hierarchically decomposed. Any particular decomposition constitutes a model of the system.[21] Klir[22] has devoted a great deal of effort to development of empirical techniques for selecting 'natural decompositions' based on methods of computer analysis.

Prior to the emergence of a paradigm there will be many conflicting models vying for primacy. Out of this emerge a particular set of models which give form to the basic system metaphor, enhance its explanatory potency, and result in a paradigm. Subsequent theorizing is for purposes of explaining new data in terms of the paradigm, reconciling anomalies between paradigm and data, and extending the scope of accepted models. In short, it is the theorizing of normal science. Likewise, further data is generated by paradigmatically sanctioned methods. For example, the Newtonian paradigm of light was that light consisted of particles. Thus any empirical method based on the idea that light was a wave phenomenon was disallowed. The effect was to stabilize the particle paradigm, allowing the consequences of this assumption to be developed.

Von Foerster[23] points out that historically paradigms have only changed after they are perfected. Only when a paradigm has been fully elaborated is it possible to recognize its limitations. In other words, a paradigm cannot be transcended until it is comprehended.

The danger inherent in the hypothetico-deductive method (Figure 2) is that by following this method a person may come to accept the paradigmatic context within which he works as fixed, dogmatic. The psycho-epistemic method specifically notes the transitory nature of paradigms. The latter are simply convenient metaphors which, by their very nature, conceal as well as reveal. For example, the particle paradigm of light allowed discovery of those aspects of the phenomenon of light which showed a particle-like nature and, simultaneously, concealed their light wave-like nature. Recognition of this idea, places the scientist in a position of having to simultaneously acknowledge both A and not A in his thinking. In modern physics this is clearly recognized in the principle of complementarity.[24]

CONCLUSIONS FOR SYSTEM SCIENCE

The recognition that there are general properties ascribable to abstract 'systems' was the beginning of system science. This has been followed by a number of proposed models of general systems.[25-27]

In terms of application in particular cases the initial step is identification of the system in question. It must be distinguished from the context in which it naturally occurs. This is followed by construction of theoretical models of the system. Each model will, in turn, suggest certain empirical methods to generate further information about the system. Once begun, this process continues automatically, at least until a satisfactory understanding of the system is achieved.

The initial distinction of a system allows choice of a set of canonical examples. These examples facilitate the process of modelling, which begins with an initial classification of system components. Naturalness of the classification scheme[28] facilitates further theorizing, and, in fact, classification and model building participate in an on-going interaction:

> '. . . classification in science is not a static thing that precedes discovery of theory but is intimately intertwined with it. Good classifications made discoveries possible, and, in turn, discoveries change our ways of classifying the things we study.'
>
> Goldstein and Goldstein[8]

Or, more metaphorically,

> 'First the taking in of scattered particulars under one idea, so that everyone understands what is being talked about. . . . Second, the separation of the idea into parts by dividing it at the joints as nature directs, not breaking any limb in half as a bad carver might.'
>
> (Plato, *Phadrus:* 265AD)

Empirical methods serve the purpose of refining classification schemes and elucidating interactions between classified system components.

An excellent example of this process is Miller's Living Systems theory.[26] Miller begins by distinguishing the set of living systems as concrete examples of a general living system. He proceeds to construct a theoretical model of this general living system, classifying seven distinct hierarchical levels, and nineteen critical subsystems within each level. In most cases he suggests specific examples. Based on this classification, and general systematic principles, he posits a number of hypotheses about living systems. Further development of his theory, according to considerations of this paper, will involve development of empirical methods to test (and modify) these hypotheses.

The model of science presented in this paper has one very pragmatic consequence. This is that the only valid criterion by which a scientific claim can ultimately be judged is success. The particle theory of light, the Ptolmic theory of planetary motion, and the caloric theory of heat are, by this criterion, valid. Each was successful in serving as a focus for the generation of operational knowledge. That each has been succeeded by other theories does not argue against their validity in terms of their purpose in the scientific enterprise, any more than the election of a public official argues against the success of his predecessors.

The picture I have tried to paint is of a threefold scientific process which generates scientific knowledge, recognizing that the continuation of this process requires paradigmatic assumptions whose very nature ensures that this knowledge is not more than a passing presence, to be judged by current usefulness rather than enshrined as dogma.

Returning to Figure 1 it should now be clear that, within the technical language of this paper, this is a model of a system of action, the system itself being the human agent acting within an environment. According to this model, the goals of such action are determined by a process involving the on-going interaction of the three sub-subsystems of the subsystem labeled mysticism, the knowledge base on which it rests is generated by the subsystem labeled science via the psycho-epistemic process of Figure 3, while the means of action are magically determined.

Note that no one of these three subsystems is ranked above the others and all three are necessary for any complete action. Nevertheless, different cultures are seen to place more emphasis on one or another. Our modern culture, for example, emphasizes the scientific and has only begun to discover the potential of magick[29] and mysticism. The medieval Islamic and Christian cultures placed strong emphasis on magick while the cultures of India and China emphasized the mystical.

REFERENCES

1. C. F. Jung, *Psychology and Alchemy* (2 rev. ed.) (R.F.C. Hull, trans.) (Princeton University Press, Princeton, N.J. 1968).
2. A. Crowley, *Magick in Theory and Practice* (New York: Dover 1976).
3. F. Capra, *The Tao of Physics* (Fontana Bungay, Suffolk 1976).
4. R. E. Ornstein, *The Physchology of Consciousness* (Viking Press, New York 1972).
5. C. Tart, (ed.) *Transpersonal Psychologies* (Harper and Row, New York 1975).
6. G. Bateson, *Mind and Nature: a Necessary Unity* New York (E.P. Dutton, New York 1979).
7. S. Grof, *Realms of the Human Unconscious* (Viking Press, New York 1975).
8. M. Goldstein and I. F. Goldstein, *How We Know* (Plenum Press, New York 1978).
9. A. Maslow, *Toward a Psychology of Being* (University Press, Princeton, N.H. 1962).
10. J. R. Royce, Three ways of knowing, and the scientific world view. *Methodology and Science* **11,** 474-505'(1978).
11. P. Feyerabend, *Against Method; Outline of an Anarchistic Theory of Knowledge* (NLB, London 1975).
12. T. Kuhn, *The Structure of Scientific Revolutions* (2nd ed. enl.) University of Chicago Press, Chicago 1970).
13. T. Kuhn, The essential tension. In: *The Essential Tension* (University of Chicago Press, Chicago 1977).
14. K. Popper, *Objective Knowledge* (Oxford University Press, Oxford 1972).
15. W. Weimer, *Notes on the Methodology of Scientific Research* (John Wiley and Sons, New York 1979).
16. J. R. Royce, How we can best advance the construction of theory in psychology, *Canadian Psychological Review* **19,** 259-276 (1978).
17. C. Tolman, Metatheoretical constructionism: a materialist evaluation, *Canadian Psychology* **21,** 7-13 (1980).
18. B. H. Voorhees and J. R. Royce, Einstein and epistemology, In: *Albert Einstein as an Interdisciplinary and Intercultural Phenomena,* N. Detlof, and L. Ugrinski, eds. (Hofstra University 1972, in press).
19. O. Ichazo, *The Human Process for Enlightenment and Freedom* (Arica Institute, New York 1976).
20. O. Ichazo, *Between Metaphysics and Protoanalysis* (Arica Institute, New York 1972).
21. B. H. Voorhees, Towards a mathematical theory of hierarchical systems in *A General Survey of Systems Methodology* (L. Troncale, ed.) (Society for General Systems Research, Louisville, KY 1982).
22. G. Klir, Identification of generative structures in empirical data, *Int. J. General Systems* **3,** No. 2, 89-104 (1976).
23. H. von Foerster, Beginnings, Ludwig von Bertalanffy Memorial Lecture, *Proceedings of Annual Meeting of the Society for General Systems Research,* Washington, D.C. (1982).

24. N. Bohr, *Atomic Theory and the Description of Nature* Cambridge University Press (Cambridge 1934).
25. M. O. Mesarovic, D. Macko and Y. Takahara, *Theory of Hierarchical Multilevel Systems* (Academic Press, New York 1970).
26. J. G. Miller, *Living Systems* (McGraw-Hill, New York 1978).
27. G. Pask, *Conversation Theory: Applications in Education and Epistemology* (Elsevier, Amsterdam 1976).
28. J. Gilmour, Taxonomy and Philosophy, In: *The New Systematics* J. S. Huxley, ed. (Clarendon Press, Oxford 1980).
29. M. Kline, *Mathematical Thought from Ancient to Modern Times* (Oxford University Press, New York 1972).

8. THE DYNAMICS OF KNOWLEDGE STRUCTURE: SEMIOTIC SYSTEMS

Maria Nowakowska

The general outline of a formal theory of semiotics is presented. Then, the methodological and integrative role of semiotics for the foundations of cognition is discussed. Finally, an application of formal semiotics to the representation of knowledge is illustrated.

INTRODUCTION

This paper presents a further extension of previously introduced theory of formal semiotics and knowledge representation.

The basic notions of the theory of formal semiotics were introduced by me at the Semiotic Colloquium in Budapest (1976) and on the Third Vienna Colloquium on Semiotics (1977), in the paper entitled 'Objects and Their Verbal Copies: Towards Foundations of Formal Semiotics'.

Further development was shown in Nowakowska[1,2] extending formal semiotics to multidimensional semiotic systems, in which the basic, or local, semiotic systems and global, or macro-level systems (composed out of local ones) were introduced. The theory of local systems was called formal semiotics.

The value of the concept of formal semiotics lies, among others, in the possibility of unification of several approaches in cognitive processes, raising some new questions, and penetrating deeper into the most crucial problems of the structure of common knowledge and belief, and finally – explicating conditions under which one may speak of 'scientific' knowledge.

As regards formal semiotics, as sketched here, it goes far beyond the notions of Peirce, leading to a system richer than taxonomical considerations, allowing for the development of semiotic theory.

On macro-levels, one considers conditions for creation of more or less integrated global semiotic systems, where the base systems cooperate in forming the overall semantics of the global system.

The formal theory of semiotics concerns three domains: (a) the domain of objects, (b) the domain of signs, and (c) the domain of meanings.

Each of these domains has its own characteristic structure, which may be represented on various levels of abstraction. In addition to that, they are related one to another by certain specific relations.

The analysis of the structure of each domain, and of the interrelations between them, constitute subsystems of the formal theory of semiotics.

GENERAL OUTLINE OF FORMAL SEMIOTICS

The starting point of the formal theory of semiotics is the relation between an object – real or imagined – identified with a certain relational structure – and its various verbal and non-verbal representations, to be called copies of the object.

The point is that objects may be represented by means of various media, and on various levels of abstraction. In other words, the same object may have various copies, built out of 'material' characteristic for the given medium.

For instance, a pictorial representation of an object, which preserves relatively rich information about it, may correspond to the notion of an icon, in the classification of Peirce.

One can also imagine pictorial representations on various levels of abstraction, up to such which preserve only the most basic features of the object, hence are very 'laconic' signs of objects. In general, icons are related to objects by some sort of similarity relation.

One can also represent objects in an arbitrary way, by assigning to them a conventional symbol, or configuration of symbols; these are symbolic signs, in Peirce's classification.

An object may also be represented verbally, that is, one can connect with it its verbal copy, again on various levels of abstraction. A special class of verbal copies are linguistic measurements, i.e. assignments of expressions from certain linguistic scales to objects or their features.

Finally, an object may be represented numerically, when some of its features are mapped into a numerical scale of a given type.

In the consideration below, all representations of objects will be generally called signs.

Formally, with each object one may connect a family of distributions determined by various representations. The point is that signs may represent objects in varying degrees, that is, in a fuzzy way. This degree of representation is expressed by a number between 0 and 1, where 1 corresponds to full representation, and 0 to no representation at all.

Thus, with each object there will be connected a function, whose arguments are signs, and the values are the degrees to which these signs represent a given object. This function is a sign potential of an object.

Since the signs are partitioned into taxonomical categories, such as verbal, iconic, etc., one can consider the representations restricted to particular categories. The union of sets of all representations of a given object by signs of a given category, forms a representation-base of this object, partitioned in a natural way into 'sub-bases'. The representation whose all sub-bases are not empty is called full, and the class of all representations in full degree (degree 1) is called the ideal of the object.

In general, one may consider the problem how well a given object may be represented with a given category of signs, and analyse various interrelations between the representations through different types of signs.

As already mentioned, a special type of signs are verbal copies, that is, representations with verbal signs. They deserve particular attention, because of the richness of their structure – which is, in effect, such as the structure of the natural language.

Other representations may also be regarded as 'languages', of a more 'local' character, in which one can distinguish their vocabularies – basic signs – and production rules, or else, restrictions and constraints on the production of composite signs. Here the main problem is the relationship between the semantics generated by the object, and mutual substitutability of the media of expression.

As mentioned, these problems concern only 'local languages', other than verbal. Here we often deal with formation of sign Gestalts, whose meaning exceeds that of a 'sum' of meanings of the components. These Gestalts appear, in turn, as components in higher order units. Here the basic operations seem to be concatenation, or conjunction, and negation.

Of special interest are the inverse images of signs; these may consist of a single element – in case of signs with unique meaning – and of many elements, in the case of signs with multiple meanings, or ambiguous. Thus, ambiguous signs allow in their inverse images other elements, perhaps with the degree of admissibility less than 1. The natural notion here is alpha-admissibility of a sign of an object, or alpha-reconstructibility of an object.

Generally, the analysis of relations between meanings of components and meaning of the whole, leads to such concepts as supporting, inhibition, cancellation, and generation of meanings.

As mentioned, the relation between objects and their signs is, by its nature, a fuzzy one. This allows the introduction of various concepts concerning the degree of expressibility of a given meaning.

If, in addition, one assumes some relation in the set of meanings describing their proximity, one obtains a possibility of explication of notions relating proximity of meanings with compositions of signs.

Signs may also be connected by other relations, such as synonymy, homonymy or antonymy. The most interesting of these relations is that of semantic implication, defined for verbal copies: A implies B, if the conjunction A *and not* B is not admissible in any sign context.

Finally, the objects themselves may be related by semantic implication, defined through the inclusion of the corresponding fuzzy sets representing signs of a given object.

There exists also another category of signs, in the classification of Peirce, namely the indexes. Here the issues concerns the causal relations between objects. When object A is a cause of object B, then B is a sign of A. In other words, one could say that object B represents object A, being a sign of it.

The relation here may have either deterministic or non-deterministic characters; it is usually complemented with explanatory sentences, concerning laws or facts. From the formal point of view, to some pairs of objects connected by causal relations, there are assigned explanatory sentences.

For each explanatory sentence in form of a law, one can speak of the set of objects which satisfy this law, that is, about the scope of the law. If such a law is denoted by P, and its domain and inverse domain are G_p and G'_p, one can speak of intersections of inverse domains, corresponding to configurations of objects. In other words, an object is described by many laws, and more precisely, by all those laws in which it appears in their inverse domain. Coherent sets of laws form theories of some sets of objects, such as situations or phenomena.

The decision whether or not a given object belongs to the domain of a given law depends, in many cases, on the observability restrictions. These restrictions lead to the fact that observing some aspects, using such or other measurement tools, may disturb and inhibit the possibilities of observing some other aspects of the object.

Thus, observation of an object is relative to the situation, to measurement procedures, and mappings in the numerical or linguistic domains, which are the prerequisites of formulating the observational judgments. The latter mapping allows us to replace the operations on objects by operations on symbols – either in mathematical domain, or in linguistic space – of classifications or descriptions. This mapping – or its character – is determined by the class of transformations of observational system, which impose constraints on its adequacy and invariance. In other words, one may deal with the whole class of descriptions or numerical assignments, from which one chooses only those which are characterized by sufficient degree of invariance. An important problem here is the choice of an adequate representation of observation of observations of objects and relations between them.

Both for the numerical representation, and for the representation in a linguistic space, the criterion of adequacy is its validity, and also parsimony – simplicity, efficiency, etc.

The system of causal relations between objects may be extended to sets of objects which constitute situations, to causal relations between networks of objects, to sets of sentences describing facts or laws, classes of procedures, numerical and linguistic mappings, and conditions for invariance.

The system so generalized allows us to introduce the generalized descriptions of objects, in terms of laws and theories.

This is a dynamical system, not only in the sense of changes of objects, and the corresponding changes of descriptions, possibilities of continuous collection of information, but also because some modifications of the observational system – for instance, due to the introduction of new types of measurement tools – or improvement of the representation for instance, by introducing a new model, induces the improved and enriched description of relations between objects. In consequence, it also leads to elimination of errors, extension of old, or introduction of new theory for the extended set of observational sentences.

To sum up, the dynamics of relations between observations and collected knowledge about objects, on the one hand, and production of their copies on a given level of representation, shows the methodological and integrative role of semiotic conceptions of foundations of cognition.

There is an especially important problem connected with the above sketched semiotic approach, namely that of optimization of operations on representations. The main question here is which cognitive processes intervene and facilitate the optimization process.

Depending on the medium and various goals of optimization, it is necessary to look for relations between cognitive processes and representations. For the social sciences, the semiotical approach offers clear and well established logical research on the structure of knowledge and its relation to reality. For philosophy, it offers a unification of ontology and epistemology. For the latter, one gets an extension of its classical scope. For logic, one gets probably a simplification and improvement of logical calculi.

For linguistics, it allows to consider a unified system lying at the base of all natural and formal languages. Also, for the foundations of informatics, the introduction of structural elements into what is described both in memory and language, may lead to new more economical principles of computer operations, and new theoretical solutions.

A THEORY OF KNOWLEDGE

This section will give an example of application of formal semiotics to representation of knowledge. We shall namely assume that the subject's knowledge about some fragment of reality is representable (on the 'first level' of representation) in the form of a set of propositions (asserting some facts, laws, conjectures, hypotheses, etc.), with each proposition accompanied possibly by a modal frame, which reflects the subject's attitude towards this proposition, e.g. 'I am almost certain that', 'I believe that', etc. (for a detailed analysis of some class of modal frames, forming motivational calculus, see Nowakowska[3]).

Concept of α-bility

We shall begin by introducing some auxiliary notions.
Let $\alpha(x,y)$ be a real-valued function defined for all $0 \leqq x,y \leqslant 1$, satisfying the following conditions:

1. SYMMETRY: $\alpha(x,y) = \alpha(y,x)$;

2. MONOTONICITY: *$\alpha(x,y)$ is increasing in x for every y*
(hence it is also increasing in y for every x). We shall actually require somewhat less, namely that α is non-decreasing in each of its arguments.

3. BOUNDARY CONDITIONS: $\alpha(x,0) = 0$, $\alpha(1,1) = 1$;

4. TRANSITIVITY: $\alpha[x,\alpha(y,z)] = \alpha[\alpha(x,y),z]$.

For a given function α satisfying the above conditions, we define its *dual*, denoted by α^*, by the relation

(1) $\alpha^*(x,y) = 1 - \alpha(1 - x, 1 - y)$.

As examples of such functions, one may take $\alpha(x,y) = \min(x,y)$ used by Zadeh in its Possibility Theory, with dual being

$$(2) \qquad \alpha^*(x,y) = \max(x,y),$$

or $\alpha(x,y) = xy$, whose dual gives probabilistic summation

$$(3) \qquad \alpha^*(x,y) = x + y - xy.$$

Somewhat more generally, we may take the functions

$$(4) \qquad \alpha(x,y) = \frac{xy}{\gamma + (1-\gamma)(x + y - xy)}, \ \gamma \geqq 0$$

or

$$(5) \qquad \alpha(x,y) = \frac{xy}{\max(x,y,\gamma)}, \ 0 \leqq \gamma \leqq 1,$$

which both reduce to $\alpha(x,y) = xy$ when $\gamma = 1$. The function (4) was introduced by Hamacher[4], while the function (5) was introduced by Dubois and Prade.[5]

As other functions, one may take, for instance

$$(6) \qquad \alpha(x,y) = \max[0, (\gamma + 1)(x + y - 1) - \gamma xy], \ \gamma \geqq -1$$

(Sugeno[6]), or

$$(7) \qquad \alpha(x,y) = 1 - \min[1, (1-x)^q + (1-y)^q]^{1/q}, \ q > 0.$$

(Yager[7]).

The dual to functions (4), (5), (6) and (7) are respectively equal

$$(8) \qquad \alpha^*(x,y) = \frac{x + y - (2-\gamma)xy}{1 - (1-\gamma)xy},$$

$$(9) \qquad \alpha^*(x,y) = \frac{x + y - xy - \min(x,y,1-\gamma)}{\max(1-x, 1-y, \gamma)},$$

$$(10) \qquad \alpha^*(x,y) = \min[1, x + y + \gamma xy], \ \gamma \geqq -1,$$

and

$$(11) \qquad \alpha^*(x,y) = \min[1, x^q + y^q]^{1/q}, \ q > 0.$$

Suppose now that we have a function $f: X \to [0,1]$, where X is some arbitrary set. For a fixed function α, the α^*-extension of f, to be denoted by f_α, will be a function

(12) $f_\alpha: 2^X \to [0,1]$

defined as follows:

(a) $f_\alpha(\emptyset) = 0$, and $f_\alpha(\{x\}) = f(x)$ for every $x \in X$;

(b) if $A = \{x_0, x_1, \ldots, x_n\}$ is a finite subset of X, then $f_\alpha(A) = c_n$, where the sequence $c_0, c_1, \ldots, c_n$ is defined inductively by $c_0 = f(x_0)$, $c_{k+1} = \alpha^*[c_k, f(x_{k+1})]$, $k = 0,1, \ldots$;

(c) for an arbitrary subset A of X, we put

(13) $f_\alpha(A) = \sup \{f_\alpha(B): B \subset A, B \text{ finite}\}$.

One may prove the following theorem.

THEOREM. *The α-extension is determined uniquely.*

Obviously, the only assertion to prove here is that condition (b) determines the constant c_n uniquely, regardless of the ordering of the set A. This property is a consequence of the assumed transitivity and symmetry of the function α.

Representation of Knowledge

Let us now consider the problems of knowledge. We shall consider a fixed subject, and assume that the knowledge of the subject may be represented as follows. There exists a set H (of all possible states, or 'histories' of relevance in the context considered), and knowledge of the subject has the form of a set of pairs

(14) $K = \{(p_i, M_i), i = 1,2, \ldots n\}$

where p_i is a certain proposition, and M_i is the modal frame, representing the observer's attitude towards the proposition p_i.

Examples of modal frames (see Nowakowska[3]) may be *I think that, I am almost certain that, I believe that, It is possible that*, etc.

With each M_i we associate a function α^* with the properties described above, while to each p_i there corresponds a family

(15 $V_i = \{H_{1,i}, H_{2,i}, \ldots, H_{n_i i}\}$

of disjoint subsets of H, and a function

(16) $f_i: V_i \to [0,1]$.

We shall now denote by V_i^* the class of all subsets of V_i and let $f_{i,\alpha}$ be the α^*-extension of f_i to V_i^*.

Finally, we denote by g_i any function mapping H into $[0,1]$, such that its α^*-extension to 2^H coincides with $f_{i,\alpha}$ on V_i^*.

DEFINITION. The knowledge K, given by (14), will be represented as a fuzzy set H_K in H, with the membership function

(17) $U_K(h) = \min g_i(h)$.

Consistency and Adequacy of Knowledge

Using the properties of the set H_K one may now define various properties of knowledge, such as its consistency. On the other hand, when one also introduces the concept of 'true' state (history) h_0, one may express such properties of knowledge as its adequacy and precision, etc.

DEFINITION. The knowledge K will be called *consistent*, if

(18) $\sup_h U_K(h) = 1$,

i.e. when the fuzzy set H_K is normal.

This means that for consistent sets H_K there exists at least one element $h \varepsilon H$ which meets all the constraints imposed by K in full degree.

On the other hand, if $U_K(h) = 0$ for all h, then the knowledge K will be called *totally inconsistent*.

Weakening the above extremes, we say that K is η-inconsistent, if $\sup U_K(h) = 1 - \eta$. For $\eta = 0$ we obtain consistency, while for $\eta = 1$ we get total inconsistency.

Even though K may be inconsistent, some of its subsets may be consistent. In particular, let

(19) $K' = \{p_{i_1}, \ldots, p_{i_j}\} \subset \{p_1, \ldots, p_n\}$.

Denote $U_{K'}(h) = \min g_{i_j}(h)$. We then say that K' is *maximally consistent*, if $\sup U_{K'}(h) = 1$, while $\sup U_{K''}(h) < 1$ for any $K'' \supset K'$ with $K'' \neq K'$.

The above concepts concerned the internal properties of knowledge K, expressed only through the properties of the fuzzy membership function $U_K(h)$.

Using now the notion of true state (or true history) h_0, we may introduce the following definition.

DEFINITION. The knowledge K is said to be *adequate*, if

(20) $U_K(h_0) \geqq U_K(h)$ for all $h \varepsilon H$

and *strictly adequate*, if

(21) $U_K(h_0) > U_K(h)$ for all $h \neq h_0$.

In particular, if $U_K(h_0) = 1$, we say that knowledge K is *truthful* (observe that truthful knowledge is always adequate, though not necessarily strictly adequate). If K is truthful and strictly adequate, it will be called *precise*.

Generally, the sets

(22) $\{h : U_K(h) = 1\}$

and

(23) $\{h : U_K(h) = 0\}$

represent the certainty in K: the first set reflects a positive and certain knowledge about the true state of the situation, while the second set – the negative and certain knowledge about the same. Consequently, the set

(24) $\{h : 0 < U_K(h) < 1\}$

may be called the *ambiguity shell* of K. It comprises that which is possible in varying positive degrees, but not certain.

REFERENCES

1. M. Nowakowska, 'Semiotic Systems and Knowledge Representation', *Int. J. Man-Machine Studies* **13**, 223–257 (1980).
2. M. Nowakowska, 'Formal Semiotics and Multidimensional Semiotic Systems', *Cybernetics and Systems: An International Journal* **12**, 83–102 (1981).
3. M. Nowakowska, *Language of Motivation and Language of Actions* (Mouton: The Hague, 1973).
4. H. Hamacher 'Uber logische Verknüpfungen unscharfer Aussagen und deren zugehörige Bewertungs-funktionen', *Progress in Cybernetics and Systems Research* (Hemisphere Publ. Co., New York, 1975), pp. 276–287.
5. D. Dubois and H. Prade, *Fuzzy Sets and Systems: Theory and Applications* (Academic Press, New York, 1980).
6. M. Sugeno, 'Theory of Fuzzy Integrals and Their Applications', *Ph.D. Thesis* (Tokyo Institute of Technology, Tokyo, 1974).
7. R. R. Yager, 'On a General Class of Fuzzy Connectives', *Fuzzy Sets and Systems* **4**, 235–242 (1980).

9. METASYSTEMS AND PROBLEMS OF FREEDOM OF CHOICE: NEW ARGUMENTS IN FAVOR OF AN OLD METAPHOR

Vladimir A. Lefebvre

A parallel between an algebraic model of ethical cognition and an idealised physical experiment showing interference of elementary particles is discussed. The indication of a slot through which a particle has passed (in the physical experiment) is analogous to an individual's choice of who is responsible for a negative situation (in the model of ethical cognition); an activated counter which registers particles passing through a given slot is similar to a correct registration of inner images by an individual. It turns out that if there are no correct registrations of inner images, it is impossible to find a function of choice of a person responsible for a negative situation. The fact corresponds to the absence of hidden parameters which would allow us to indicate a slot through which a particle has passed if the counters are non-activated. Thus, an individual's choice of a person responsible for a negative situation cannot be described functionally if the individual does not have correct registrations of his inner images. If we call such a choice 'a freedom of will', then the concepts 'freedom of will' and indeterminism would be equal.

INTRODUCTION

Among the various philosophical reactions to the appearance of quantum mechanics, one was a raft of speculations of the topic of analogies between the conscious actions of man and the behaviour of elementary particles (see the analysis of this kind of problem in the book *Philosophy of Science* by Frank.[1]). On the one hand, in the peculiarities of the behaviour of electrons, philosophers saw a manifestation of the phenomenon of 'freedom of will' similar to free human choice; on the other hand, in freedom of choice some scholars saw a manifestation of physical indeterminism.

Numerous discussions on the topic have produced no new or constructive ideas. Study of the problem of choice in psychology has developed independently of quantum mechanics, and physicists today are inclined to consider analogies between the two fields as curious metaphors. (There are, however, some exceptions: in his works Orlov[2,3] developed a 'wave logic'; using quantum-mechanical concepts he constructed a model of an individual's 'free choice' of the meaning of a logical proposition; the act of this choice is considered analogous to a physical measurement. Gurevich and Feigenberg[4] in their paper discuss the nature of uncertainty in human choice.)

The present author chanced on this analogy quite unexpectedly while analyzing his algebraic model of ethical cognition, a model constructed without any connection to physical conceptions. At the basis of the model there is a representation of the human inner world as a tree-like hierarchy of meta-images.[5–10]

It turned out to be possible to analyze human inner feelings and the knowledge about these feelings in a single model. And correlations similar to the principle of complementarity of Niels Bohr were found in this model. (Bohr[11] suggested that correlations with complementarity may be realized in the inner world of man.) In particular, it turned out that with the help of the algebraic model of ethical cognition it is possible to model the famous idealized experiment illustrating the interference of elementary particles which pass through two parallel slots onto a screen.

It is known from physics that the interference appears only when an observer cannot find out *even potentially* through which slot a particle has passed. In the opposite case the interference disappears, and an observer sees on the screen the sum of the usual statistical distributions from each slot.

Something similar can be observed in the algebraic model, if we have the two physical slots correspond to an 'image of oneself' and an 'image of one's partner', and have the various statistical distributions of electrons on the screen correspond to 'feeling of guilt', 'feeling of condemnation', and 'feeling of suffering about a negative situation'. Determination of a slot through which a particle has passed corresponds to the indication of an individual for the appearance of a negative situation. It turns out that in the case where in the algebraic model of ethical cognition there is an analogue of the physical interference, an individual cannot indicate who is responsible for a negative situation: he or his partner. And this uncertainty is fundamental: we cannot extend any function of choice (with the properties 'natural' for the functions of choice) to this case.

1. A BRIEF DESCRIPTION OF THE ALGEBRAIC MODEL OF ETHICAL COGNITION

The essence of the model is as follows. An individual is represented by an abstract automaton

$$\{x_1, x_2, y, \phi, A\}$$

where

x_1 and x_2 are inputs;

y is an output (x_1, x_2, y have values of Boolean 1 or 0);

ϕ is a Boolean function transforming the values of the inputs into the value of the output;

A is a special formula expressing the Boolean function.

x_1 and x_2 are interpreted as demands of the 'social environment' to accomplish 'good' or 'bad' actions (respectively to Boolean values 1 and 0); y is interpreted as a 'good' (1) or 'bad' (0) individual's response. If the frequencies of 0's at the inputs are given, we can calculate the frequency of 0's at the output, i.e., the frequency of an individual's 'bad' reactions.

The main novelty of the model is the representation of a transforming function

ϕ with a special formula A: *its syntax allows us to describe an individual's inner world* utilizing such categories as image of oneself, image of one's partner, feeling of guilt, condemnation, and suffering. *Our model relates two different approaches by which the representation of human behaviour can be made: the stimulus-response approach and the inner cognition approach.*

For a better understanding of this idea let us look at Figure 1. In the head of a cartoon man we are inserting images of himself and his partner which in turn have images of himself and the partner. The picture can be coded, one to one, to the hierarchical formula given at the bottom of Figure 1. This formula is a regular logical expression, but written in an exponential form..The exponent corresponds to implication. In the traditional notation this expression would

$$(((\mathrm{a} \cdot \mathrm{b}) \supset a\ ) \cdot ((\mathrm{a} \cdot \mathrm{b}) \supset b))) \supset a,$$

where a is a or $\bar{a}$, and b is b or $\bar{b}$. In this way we can equate this kind of picture with Boolean functions.

Let us consider in more detail the fragments constituting the representation of an individual (Figure 2 upper description). We shall call elements A_1 and B_1 the images of oneself and of one's partner respectively, and elements a and b the registering meta-images or simply the registrations. Registrations $\mathrm{a} \equiv a$ and $\mathrm{b} \equiv b$ are called correct, and registrations $\mathrm{a} \equiv \bar{a}$ and $\mathrm{b} \equiv \bar{b}$ are incorrect. The frequency of the appearance of zeros at the output of some fragment M of individual A will be designated as $\|M\|$. We will use the following interpretations:

$\|A_1\|$ – feeling of guilt

$\|B_1\|$ – feeling of condemnation of a partner

$\|A_1 \cdot B_1\|$ – feeling of suffering about a negative situation.

A more detailed description and motivation of such interpretations are given in the book *Algebra of Conscience*[10].

$$a^{a^{\mathrm{a}\cdot\mathrm{b}}\cdot\, b^{\mathrm{a}\cdot\mathrm{b}}}$$

Fig. 1. Reflexive structure and the corresponding algebraic expression.

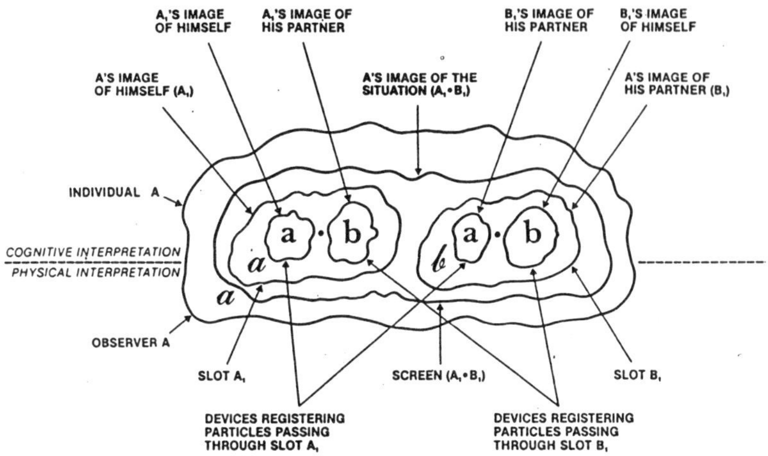

Fig. 2. Two interpretations of the algebraic expression: cognitive interpretation in the upper part and physical interpretation in the lower part.

2. PARALLEL: THE INTERFERENCE OF FEELINGS AND THE INTERFERENCE OF ELEMENTARY PARTICLES

We have demonstrated that the inner world of individual A can be put into correspondence with the experimental situation where interference is observed (Figure 2, lower description). We will put correct registrations into correspondence with activated counters, and incorrect with non-activated. The special theorem has been proven that if individual A has no correct registrations at all, then the analogue of interference takes place:[10]

$$\|A_1 \cdot B_1\| \neq \|A_1\| + \|B_1\|.$$

If there is at least one correct registration, then there is no analogue of interference:

$$\|A_1 \cdot B_1\| = \|A_1\| + \|B_1\|$$

Therefore, both in the algebraic model of ethical cognition and in the physical experiment with two slots, the disturbance of arithmetic additivity (interference) disappears, if there is a correct registration.

In the physical experiment a correct registration allows the investigator to determine which slot a particle has passed through, and in the algebraic model of ethical cognition it allows the individual to indicate who is responsible for a negative situation (A_1 or B_1) at a given moment.

3. THE MECHANISM OF CHOICE

Let us analyze the case where individual A has two correct registrations:

$$a \underbrace{\bar{a} \cdot \boxed{b}}_{A_1} \qquad b \underbrace{\boxed{a} \cdot \bar{b}}_{B_1}$$

Correct registrations are framed. Consider table I:

Table I
Values of various parts of an individual who has two correct registrations

	1	2	3	4	5	6	7
	a	b	A_1	B_1	$\boxed{a}$	$\boxed{b}$	$A_1 \; B_1$
1	1	1	1	1	1	1	1
2	0	1	0	1	0	1	0
3	1	0	1	0	1	0	0
4	0	0	1	1	0	0	1
frequencies			$\lVert A_1 \rVert = \frac{1}{4}$	$\lVert B_1 \rVert = \frac{1}{4}$			$\lVert A_1 \cdot B_1 \rVert = \frac{1}{2}$

In columns 1 and 2 various combinations of input variables are given. Columns 3 and 4 indicate the corresponding outputs of individuals A_1 and B_1. At the very bottom there are frequencies of the appearance of zero. Columns 5 and 6 show the values of correct registrations, and column 7 indicates the 'value of a situation'. We see that the frequency of its becoming zero is equal to the sum of frequencies of becoming zero of A_1 and B_1.

Consider the case where individual A sees a negative situation. Negative situations correspond to lines 2 and 3. Individual A *feels* either an 'impulse of guilt' ($A_1 = 0$), or an 'impulse of condemnation' ($B_1 = 0$); in addition he *knows* that he feels an impulse since simultaneously with experiencing an impulse the corresponding registration also becomes zero. Thus, in this case an individual observes a negative situation, feels some impulse, and knows who is responsible for the negative situation: he (line 2) or his partner (line 3).

It is easy to show that if an individual has at least one correct registration and sees a negative situation ($A_1 \cdot B_1 = 0$), he experiences either an impulse of guilt ($A_1 = 0$) *or* an impulse of condemnation ($B_1 = 0$). And if in addition neither A_1 nor B_1 is equal to 1 identically (i.e., an individual 'sometimes' experiences an impulse of guilt and sometimes an impulse of condemnation), then a correct registration $\boxed{a}$ (if it exists) is equal to 0 only when $A_1 = 0$, and a correct registration $\boxed{b}$ (if it exists) is equal to 0 only when $B_1 = 0$. Thus an individual has a natural function of choice of a person responsible for a negative situation, and this function is based only on his knowledge of values of the correct registrations. A description of the function of choice of an individual who has at least one correct registration under conditions that $A_1 \cdot B_1 = 0$ and both $A_1 \neq 1$ and $B_1 \neq 1$ is as follows:

(1) There are two correct registrations $\boxed{a}$ and $\boxed{b}$. Then, if $\boxed{a} = 0$, an individual considers himself responsible for a negative situation, and if $\boxed{b} = 0$, an individual considers his partner to be responsible. (This is the case analyzed above.)

(2) There is only one correct registration – $\boxed{a}$. Then, if $\boxed{a} = 0$, an individual considers himself responsible for a negative situation, and if $\boxed{a} \neq 0$, he considers his partner to be responsible.

(3) There is only one correct registration – $\boxed{b}$. Then, if $\boxed{b} \neq 0$, and individual considers himself responsible for a negative situation, and if $\boxed{b} = 0$, an individual considers his partner to be responsible.

It is easy to see that this function of choice of a person responsible for a negative situation coincides with the procedure of determination of which slot a particle has passed through (and reached the screen). The value $A_1 \cdot B_1 = 0$ corresponds to reaching the screen, and the values $\boxed{a} = 0$ and $\boxed{b} = 0$ correspond to the registrations of a particle by activated counters: the value $\boxed{a} = 0$ means that a particle has passed through slot A_1, and the value $\boxed{b} = 0$ indicates that a particle passed through slot B_1. As we have mentioned earlier, in this case there is no interference:

$$\|A_1 \cdot B_1\| = \|A_1\| + \|B_1\|$$

In the case where an individual has no correct registrations at all, he cannot make a choice using values $\boxed{a}$ and $\boxed{b}$.It is natural to suppose that an individual has some other function which allows him to determine a person responsible for a negative situation.

We will call individuals A_1 and B_1 symmetrical if they transform into each other when a is changing to b and b to a.

Let us formally describe now a function of choice of an individual responsible for a negative situation:

(1) f is a simple function of pairs of values (a,b).

(2) f has values either 1 or 2; if $f = 1$, A_1 is responsible; if $f = 2$, B_1 is responsible.

(3) Let A_1 and B_1 be symmetrical individuals; if there is a pair $a = \alpha_1$ and $b = \beta_1$ such that $f = 1$, then there is another pair $a = \alpha_2$ and $b = \beta_2$ such that $f = 2$.

Condition (3) means that there is no case where only one individual is always responsible for the negative situation. We require the choice to be alternate.

Consider now individual A who has no correct registrations.

$$\underbrace{a^{a^{\bar{a} \cdot \bar{b}}}}_{A_1} \cdot \underbrace{b^{\bar{a} \cdot \bar{b}}}_{B_1}$$

This individual corresponds to Table II.

Table II
Values of various parts of an individual who has no correct registrations

	1	2	3	4	5
	a	b	A_1	B_1	$A_1 \cdot B_1$
1	1	1	1	1	1
2	0	1	1	1	1
3	1	0	1	1	1
4	0	0	0	0	0
frequencies			$\|A_1\| = \frac{1}{4}$	$\|B_1\| = \frac{1}{4}$	$\|A_1 \cdot B_1\| = \frac{1}{4}$

In this case, when individual A sees a negative situation, he simultaneously feels an 'impulse of guilt' ($A_1 = 0$) and an 'impulse of condemnation' ($B_1 = 0$), but he 'does not know' this, since he has no correct registrations at the meta-level. Moreover, if we inform individual A about the values of the arguments, he will not be able to make the alternate choice, since $A_1 \cdot B_1 = 0$ only when one pair of arguments $a = 0$ and $b = 0$. In this case *function f dies not exist*, since A_1 and B_1 are symmetrical individuals and there is only one pair of values (a,b) which makes $A_1 \cdot B_1$ equal to zero. Therefore, the set of values of arguments is too poor to allow the construction of function f of alternate choice.

This situation is similar to the case where all counters registering particles passing through slots are non-activated. First, the interference appears:

$$\|A_1 \cdot B_1\| \neq \|A_1\| + \|B_1\|.$$

Second, we can see here 'the absence of hidden parameters,' similar to that established by von Neumann's theorem in quantum mechanics. In our model a and b might be hidden parameters. However, even knowing the values of these parameters an individual cannot indicate the slot through which a particle has passed.

If we suppose now that we observe alternate choices made by an individual without correct registrations, we can state that these alternate choices cannot be considered as a function of input stimuli a and b. In this case we might assume that we observe a manifestation of the 'freedom of will' of an individual. 'Freedom of will' is the choice of the alternative under the condition that the function of choice does not exist.

CONCLUSION

If we suppose that each electron, when no one observes it, is still passing through a certain slot, then according to von Neumann's theorem about the absence of hidden variables, we have to accept that we cannot restore the choice of an electron using deduction. An analogical situation is to be found in the alegebraic model of ethical cognition. If an individual without correct registrations is still making the alternate choices of a person responsible for a negative situation, then we have to accept that we cannot restore his choices with the help of deduction, knowing input parameters *a* and *b*.

In both cases we can speak (depending on our taste) either about 'freedom of will' or about 'indeterminate behaviour'. The old metaphor about an electron's freedom of will is acquiring the new modelling sense.

REFERENCES

1. P. Frank, *Philosophy of Science* (Prentice-Hall, Inc. Englewood Cliffs, N.J. 1957).
2. Yuri F. Orlov, 'A Quantum Model of Doubt', *Annals of the New York Academy of Sciences* **373**, 84–92 (1981).
3. Yuri F. Orlov, 'The Wave Logic of Consciousness: A Hypothesis' *Int. J. Theoretical Physics* **21**, No. 1, 37–53 (1982).
4. I. I. Gurevich and I. M. Feigenberg, 'Kakie Veroyatnosti "rabotaut" v Psichologii' (What Kind of Probabilities 'Work' in Psychology), I. M. Feigenberg and G. E. Zhuravlev, Eds. *Veroyatnostnoe Prognozirovanie Deyatelnosti Cheloveka* (Moscow, Nauka, 1977).
5. V. A. Lefebvre, *Konfliktuyushchie Struktury* (*Conflicting Structures*) (Moscow, Vysshaya Shkola, 1967).
6. V. A. Lefebvre, 'A formal Method of Investigating Reflective Processes' *General Systems* **XVII**, 181–188 (1972).
7. V. A. Lefebvre, 'A Formal Approach to the Problem of Good and Evil', *General Systems* **XXII**, 183–185 (1977).
8. V. A. Lefebvre, *The Structure of Awareness: Toward a Symbolic Language of Human Reflexion* (Sage Publications, Beverly Hills, 1977).
9. V. A. Lefebvre, 'An Algebraic Model of Ethical Cognition', *J. Mathematical Psychology* **22**, 83–120 (1980).
10. V. A. Lefebvre, *Algebra of Conscience* (Dordrecht, Holland, D. Reidel Publishing Co. 1982).
11. N. Bohr, *Atomic Physics and Human Knowledge* (John Wiley, New York, 1958).

10. INFORMATION HOLISM: TOWARD AN INTEGRATION OF THE HOLOGRAPHIC AND ANALYTIC MODELS OF CONSCIOUSNESS

John R. Battista

The holographic and analytic models of consciousness are shown to be integrated by the theory of informational holism in which consciousness is equated with information. The holographic and analytic models represent the different means by which the brain generates information. The role of holographic and analytic mechanisms in sensation, perception, emotion, cognition, intuition and unition is delineated as part of a hierarchical information theory of consciousness. The information theory of consciousness is also shown to extend the holographic and analytic models by predicting the conditions under which conscious experience takes place. Finally, informational holism is shown to clarify the relationship between brain and consciousness and make predictions about consciousness in non-human systems. Overall informational holism is revealed to be a theory of consciousness which can explain the conditions under which consciousness occurs, while the analytic and holographic models represent models of consciousness which can explain how consciousness develops in a particular situation.

THE HOLOGRAPHIC AND ANALYTICAL MODELS

We are experiencing a growing recognition of the limitations of the analytic model based on the digital computer. Quantum physics has led us to recognize the interconnectedness of all events and psychology has become increasingly aware of the inability of the analytic model to account for the intuitive and transpersonal dimensions of consciousness.

The work of Karl Pribram[1-3] and David Bohm[4,5] has generated a rapidly growing interest in the application of holographic concepts to our understanding of consciousness and the universe. Thus a new holographic model is being developed which emphasizes the interdependent, parallel and simultaneous processing of events.

We need to avoid making the same mistake with the holographic model as we did with the analytic model—to try and explain everything with it. That kind of approach would make little sense because many phenomena (like telephone communication) seem best conceptualized analytically while other phenomena (like transpersonal states of consciousness) seem best understood holographically.

Pribram[6] uses the holographic and analytic models in this complementary sense. Basically, he believes that the brain functions both as an analytic (digital) and holographic (analog) computer. He accepts that the brain functions analytically in the sense that data are processed sequentially according to a set of specific programs, the output of which is fed-back into input for the purpose of comparison. For example, the temporal cortex effects vision by programming the functions of the primary visual projection system. This allows him to explain data showing that lesions to the inferior temporal cortex will greatly impair the ability to make visual discriminations. Similarly, such a model could explain motivated perception, as when a haematologist who is scanning a bone marrow for cancer cells 'sees' only the cancerous cells. Alternately, Pribram argues that the brain is holographic, in that data are processed as a whole, as the result of interactions occurring in and between synaptic clefts. This allows him to explain the image-like nature of perception and the ability to maintain sensory capacities, despite massive insults to the sensory system. Thus, it is Pribram's use of both the holographic and analytic models that has allowed him to develop a more encompassing and integrating theory of the brain than those of theorists who have exclusively employed an analytic model.

However, if the analytic and holographic models are complementary rather than competitive this implies that there may be a more encompassing theoretical structure that is capable of integrating both of these models. My attempt to discover this structure led me to examine the basic assumptions which underly both the analytic and holographic models. This examination revealed that both the analytic and holographic models are based on a set of holistic assumptions, and led me to conclude that information theory is a theoretical structure capable of integrating both of them.

All models and theories are based on assumptions. I use the term paradigm to refer to the set of these underlying, and often recognized, assumptions. Paradigms can be differentiated by the way in which they answer the three basic questions of the philosophy of science:

1) What *is*? or what is the nature of reality?
2) How do we know what *is*? or what constitutes knowledge?
3) What accounts for change and stability in what *is*?

The answer to the first question, constitutes the paradigm's ontology; to the second, its epistemology; to the third, its dynamics or means or explanation. Table 1 provides a comparison of the three main paradigms of Western thought as I view them: vitalistic, mechanistic, and holistic. Our discussion will be limited to a consideration of the holistic paradigm but the interested reader is referred to a previous article[7] for a fuller discussion of the mechanistic and vitalistic paradigms.

According to the holistic paradigm the entire universe is interconnected and hierarchically organized. Matter and energy, living and nonliving, mind, body, and spirit all refer to different levels of the same, unified system. We know about this universal system because of our interaction with it. Uncertainty is an inherent part of our relationship to the system because the system is a process and we are part of the process we are attempting to know. To the extent we can reduce this

Table 1. *Vitalistic, Mechanistic and Holistic Paradigms*

Parameters	*Vitalism*	*Mechanism*	*Holism*
Ontology	Dualistic	Dualistic	Monistic
Epistemology	Subjective	Objective	Interactive
Methodology	Phenomenological	Empirical	Analogical
Causality	Teleological	Deterministic	Probabilistic
Analysis	Metaphysical	Reductivistic	Structural
Dynamics	Nullentropic	Entropic	Negentropic

uncertainty, we generate the information about the world process. Knowledge thus exists in relationship, rather than in an 'objective' world or in 'subjective' experience. The world is not classically determined in a simple, linear billiard ball manner, rather, each level of the universal system operates on every other. Events are thus partially determined from above (more complex and encompassing levels) and from below (less complex and encompassing levels) as well as from the same level of complexity. Thus, we can never know the absolute cause of anything, nor whether an event is totally determined or not.

In a general way, these assumptions have received a wide acceptance. In fact, it seems fairly clear that we are now in the process of a paradigmatic shift in which all fields are becoming revisioned in light of these assumptions.[8–10]

However, it may not be so clear why I believe that both the holographic and analytic models are holistic. I imagine some people may think that the holographic model and holistic paradigms are synonymous, and that the analytic model is really an example of the now outdated mechanistic paradigm from which we need to free ourselves.

It seems to me that the key to grasping the holistic and complementary nature of the holographic and analytic models lies in an understanding of the concept of *information*. Although information is commonly thought of as an entity or thing carried by a signal or input, this mechanistic interpretation is actually outdated. In information theory 'information' is defined in terms of a *relationship* between an input and a receiving device. From this perspective, it is impossible to talk about information independent of some receiver. A book contains no information for someone who cannot read. It is this relational aspect of information that makes it a holistic rather than mechanistic concept. An input is said to have carried a certain amount of information when it reduces the uncertainty associated with the receiving or measuring device. The selective value of the input refers to the *meaning* of the information, while the *amount* of information is a measure of the degree of uncertainty reduction that results from the input.[11]

For example, if a man is conversing with a woman and wishes to know if she is married or not and then notices a diamond ring on the 4th finger of her left hand and concludes that she is married, the *meaning* of the ring—its 'selective value' to the man—is that she is married. The *amount* of information carried by the perception depends on the degree of uncertainty that the man initially had about whether or not the woman was married. The uncertainty of a finite set (H) is represented mathematically[12] as

$$H = -\sum_{R=1}^{n} p_R \log_2 p_R$$

In this case there are two possible alternatives—married or unmarried, and the probability of whether the man believes the woman to be married will determine the amount of information carried by his perception of the ring. Thus the perception of the ring will carry more information if the man thought the probability the woman was married was 50% than if he was already 90% sure she was married. In the first case, he is largely uncertain as to the woman's status, and thus the perception of the ring greatly reduced his uncertainty—that is why 'information' and 'reduction of uncertainty' are synonymous. In the second case, he is not in a state of very much 'uncertainty', so the ring carried much less information for him.

What does this concept of information have to do with the analytic and holographic models? The essential point is that both the analytic and holographic models refer to the processes for generating information—and information itself is a *holographic concept*, as I explained. In the analytic model, information results, from the selection of a particular outcome from a set of possibilities. In the holographic model, information results from an analog, mechanism that allows two states to interact with each other. Thus, although the analytic and holographic models refer to distinct means of generating information, they both rest on holistic assumptions and can be considered as complementary holistic models. Furthermore, because the analytic and holographic models both refer to the generation of information, information theory is revealed to be an underlying structure that can integrate both the analytic and holographic models. It is interesting to note in this regard that Gabor, the inventor of the hologram, was also one of the three independent originators of the concept of information.[13]

This conclusion about the relationship between information theory and the holographic analytic models is important because if we accept Pribram's view that the brain functions both holographically and analytically, it implies that a general theory of consciousness needs to be based on holistic assumptions and utilize information theory. I will thus attempt to show how an information theory of consciousness can integrate and extend the contributions of the analytic and holographic models of consciousness.

AN INFORMATION THEORY OF CONSCIOUSNESS

The particular information theory of consciousness to be utilized[14] rests on three basic concepts:

1) Consciousness is information.
2) The different forms of consciousness refer to different hierarchical levels of information.
3) The intensity of consciousness at any level is a function of the amount of information at that level.

First, this information theory of consciousness can integrate the holographic and analytic models of specific forms of conscious experience into a general theory of consciousness by providing a hierarchical structure which can relate different forms of consciousness to each other.

We have learned from the pioneering work of Tart[15] that the problem of consciousness is not a unitary one but must encompass each specific state of consciousness. Although there is no standard nomenclature of states of consciousness any complete theory of consciousness must include sensation, perception, emotion, cognition, intuition, self-awareness, and unition (transpersonal consciousness).

I have little doubt that we will find both the holographic and analytic models useful in understanding these states. For example, the sensory system appears to be analytic in character in that each sensory neuron responds to a particular frequency pattern in the environment and translates that pattern into a sequential rate of neuronal discharges. Alternately, perception, as mentioned previously, may involve both holographic and analytic processes. The situation with emotions remains unclear, but the well-known work of Schacter and Singer[16], showing that a particular state of arousal is compatible with a number of different emotions depending on the environmental circumstances under which it occurs, suggests that some analytic mechanism is involved. Work on cognition and intuition is commonly conceptualized in terms of two different information processing structures, one found predominantly in the dominant hemisphere, the other in the non-dominant hemisphere. The dominant hemisphere structure processes information in an analytic (linear, sequential) manner and is thought to be involved in cognition. The non-dominant hemisphere may process information in a holographic (holistic, transformational) manner and is thought to be involved in intuition and dreaming.

Anderson[17] and Bentov[18] have both made excellent cases for the role of a holographic mechanism in transpersonal states. Basically, they contend that all of the potential information about the universe is holographically encoded in the spectrum of frequency patterns that constantly bombard us. Through meditation one quiets the brain so that it can become sympathetically in tune with (entrained to) this universal frequency pattern. When this occurs, the encoded information about the universe becomes holographically decoded, and the individual experiences a state of unitive consciousness with the entire universe. I find this model appealing for three reasons. First, the common experience of having a deeper meditation with a group of experienced meditators is compatible with this model because such a group could set up a more powerful localized field with which to become entrained. Second, the EEG findings of a number of investigators[19–21] showing the synchronization of the entire cerebral cortex during such states lends support to the idea of a holographic mechanism involving the entire brain. Third, such a model could explain the experience of knowing everything at once in such unitive states.

The value of an information theory approach is that it can integrate each of these specific models of consciousness into an integrated general theory of consciousness. This is possible because the holographic and analytic mechanisms can be understood as devices for generating information and the different states of conscious experience can be related to each other in terms of hierarchical levels of information. Table 2 provides an outline of the relationship between specific states of consciousness and levels of information. This relationship is more fully discussed in a previous paper.[22]

Second, an information theory of consciousness can extend the holographic and analytic models of consciousness by explaining the conditions under which conscious experience occurs. For example, the well known phenomenon of habituation is not readily explained by either the holographic or analytic models of perception. The habituation of perception reveals that we only perceive a phenomenon when there is an alteration in the sensory input associated with it, i.e. if a sensory input continues unaltered it will cease to be consciously perceived until it is

Table 2. Information and Consciousness

State of Consciousness	Level of Information
Sensation	*Information 1*
Perception	*Information 2*, i.e., information about information 1 (the meaning of information).
Emotion	*Information 3*, i.e., information about information 2 (the meaning of perception).
Awareness	
a. Cognition	*Information 4a**, information about information levels 2, 3, 4, 5, 6, 7 (reflective knowledge of the other forms of awareness).
b. Intuition	*Information 4b**, information about information levels 2, 3, 4, 5, 6 (non-reflective knowledge of the other forms of awareness).
Self-awareness	*Information 5*, information about information 4 and 6 (knowing the nature of one's own awareness).
Unition	*Information 6*, information about information 5 and 7 (experience of the process of awareness itself).
Absolute	*Information 7*, an integrated awareness of all the levels of awareness (pure awareness).

*Information 4a and 4b are two different types of information 4 that result from the two different structures by which this information is generated. Information 4a results from the structure of sequential logic, i.e., the Klein group, and refers to the kind of information processing pre-dominantly associated with the dominant hemisphere and which is involved in cognition. Information 4b results from the structure of the transformational logic, i.e., the transformation group, and refers to the kind of information processing predominantly associated with the non-dominant hemisphere and which is involved in intuition, dreaming, and some preliminary stages of meditation, etc. See Battista[14] for a somewhat fuller discussion of these ideas based on the work of Saul Paul Sirac.

altered or stopped. According to a holographic model of perception a holographic perception will be formed whenever a sensory input matches a standing wave in the junctional microstructure because a holographic wave (convolutional integral) has the property of decoding to produce a hologram whenever it interacts with itself. Thus this model would predict the occurrence of perception whenever the input remains constant over time-just the opposite of what actually occurs.

This situation is clarified by an information theory of perception. According to this theory, perception is a second level of information that refers to information about sensations. The theory states that the intensity of perceptual consciousness will be a function of the amount of this second level of information that is generated by a given sensory input to the perceptual system. When a new sensory input occurs it is novel and unexpected, generates a relatively large amount of information and will thus tend to be conscious. However, with repeated stimulation its novelty wears off, its probability of occurrence increases and the amount of information it carries decreases so that it will tend to become unconscious. When the stimulation is fully expected, the failure of the stimulus to occur is now unexpected, has a low probability of occurrence, carries a relatively large amount of information, and will result in perceptual awareness. A previous paper,[22] discusses the ability of an information theory of consciousness to explain the conditions under which cognition and dreaming take place. Thus, an information theory of consciousness is able to extend the value of holographic and analytic models of consciousness by explaining the activities under which various forms of consciousness occur.

Third, an information theory of consciousness can extend the value of the holographic and analytic models of consciousness by more clearly explaining the relationship between consciousness and brain states. Whenever I read an article on the physical basis of consciousness using analytic or holographic terms I always feel like asking the question—what is consciousness? It seems to me analytic theorists imply consciousness is a brain state while holographic theorists imply consciousness results from the interaction of a brain state. The relationship between brain states and consciousness becomes clearer if we accept that consciousness is information. Information is a measure of uncertainty reduction. Uncertainty reduction in the brain takes place by the interaction of sets of neurons so that a resulting brain state represents a certain amount of information.

Thus this position is consistent with the interactionist model of consciousness employed by holographic theorists because information is not a thing, but a relationship. Alternately, this position supports the analytic model of consciousness in so far as it implies consciousness is represented by a brain state. However, it disagrees with the analytic model of consciousness in so far as it implies consciousness can be reduced to a brain state. This is false because the amount and meaning of information carried by a particular brain state can vary over time. The same brain state can carry different amounts of information at different points in time because the amount of information it carries is a function of the probability of distribution of its occurrence relative to all of the other brain states, a distribution that can be expected to vary through time. Further, the *meaning* of the information can vary through time because particular brain states may be reprogrammed to take on different functions as the result of feedback from experience. Thus, consciousness may be carried by brain states but is not reducible to them. This conclusion provides a theoretical explanation for Sperry's[23,24] position that consciousness is a higher-order, molar property of the brain that exists in the same relationship to neuronal events as water does to the atoms that comprise it.

Fourth, an information theory of consciousness extends the holographic and analytical models of consciousness because it is able to make predictions about the occurrence of consciousness in non-humans. The occurrence of consciousness in non-humans is left unclear by analytic and holographic models of consciousness because it has not been clarified if consciousness is solely a property of human nervous tissue. (Bentov's[25] work is a notable exception to this generalization.) With an information theory of consciousness, this confusion is resolved: anything that generates information, whether holographically or analytically, is predicted to have consciousness.

This prediction has some very remarkable implications. First, all physical interactions are predicted to have basic sensory consciousness (info 1) because all quantum events reduce the uncertainty inherent in the Schrödinger wave equation by their occurrence. However, the degree of sensory consciousness is predicted to be a function of the amount of information represented by the state of a particular object. It can thus be seen that inanimate objects such as rocks which alter their states very little under a wide variety of environmental circumstances will extract much less sensory information from the environment than living processes such as human beings which are capable of distinguishing a large number of environmental states. Second, a wide variety of non-human objects would be predicted to have perceptual consciousness. For example, computers would be expected to be capable of perception because they can generate a second level of information (information about information). Third, an information theory of consciousness gives support to the idea of the consciousness of groups and organizations which transcend traditional physical boundaries and are defined by their structures. Most importantly in this regard, the universe as a whole can be conceptualized as a process entailing a seventh level of information that refers to pure awareness (see Table 2). Such a possibility gives support to the existence of transpersonal (spiritual) dimensions of consciousness which are independent of our awareness of them.

TOWARDS A GENERAL HOLISTIC THEORY

It is now possible to review the essential points in this paper from a somewhat different perspective. Basically a distinction has been made between scientific models, paradigms and theories. A scientific model is an analogy which attempts to explain one thing in terms of another. Theories rely on a set of abstract constructs which are used to represent and explain phenomena by the development of a set of verifiable hypotheses about them. Paradigms refer to the set of implicit basic assumptions which underlie models or theories.

There has been increasing dissatisfaction in the 20th century with the set of mechanistic assumptions which underlie our traditional scientific theories such as Newtonian physics. The first holistic model to be developed (the analytic model) was based on the digital computer and emphasized the independent, linear, sequential processing of information. This holistic model seems to have arisen first because it bears some resemblence to the old mechanistic energy drive models utilized in traditional physics. However, this model is actually holistic in that it utilizes concepts of information, uncertainty, organization, probability and entropy rather than energy, drive, force, mass, and determination. As this analytic model became more and more widely applied, its limited applicability became more and more apparent. In order to explain those phenomena, not well explained by the analytic model (e.g. the quantum interconnectedness of the universe, the panoramic nature of perception, the existence of transpersonal states of awareness) a new holistic model is emerging. This new model, the holographic, stresses the interdependent, parallel and simultaneous processing of information.

Confusion has developed because it has been unclear that both the analytic and holographic models are holistic. Thus, there has been the tendency to see the holographic and analytic models as competitive rather than complementary.

However, both the analytic and holographic models are part of the emerging development of a general holistic theory based on information theory. Information theory can explain *what* is happening while the analytic and holographic models explain *how* it is occurring.

Therefore formal holistic theories need to be constructed using information-systems concepts. As this occurs the holographic and analytic models can hopefully be integrated into a richer and more encompassing understanding of the phenomenon under consideration. This paper has attempted to provide an example of this by showing how an information theory of consciousness is capable of integrating both the analytic and holographic aspects of brain functioning. It was further implied that conceptualizing consciousness in terms of information can result in some extensions of our understanding of consciousness beyond what can be realized using holographic or analytic models alone.

Thus, the overall value of the holographic model lies in its ability to complement the analytic model and reveal to us a new means by which information can be generated. This model will have wide applicability and certainly may have great importance in helping us to explain the universe as a whole, perception, intuition, memory, unition, etc. However, the limitations of the holographic model lies in the fact that it is a model rather than a theory. The theory of holograms lies in information theory and information theory is needed to explain and predict *what* is occurring, much as the holographic and analytic models are needed to explain *how* something is occurring.

ACKNOWLEDGEMENT

My thanks to Ken Wilber for his stimulating and helpful review of this work.

REFERENCES

1. K. Pribram, *Languages of the Brain*, (Prentice Hall, Englewood Cliffs, N.J., 1971).
2. K. Pribram, "Towards a Holonomic Theory of Perception", In: S. Ertel (Ed.), *Gestalt theorie in der Modernen Psychologie*, (Steinkopff, Darmstadt, 1975).
3. K. Pribram, "Problems Concerning the Structure of Consciousness" In: G. Globus, (Ed.), *Consciousness and the Brain*, (Plenum Press, 1976).
4. D. Bohm, "Quantum Theory as an Indication of a New Order in Physics, Part A, The Development of New Orders as Shown Through the History of Physics", *Foundations of Physics*, **1**, 359–381, (1971).
5. D. Bohm, "Quantum Theory as an Indication of a New Order in Physics, Part B, Implicate and Explicate Order in Physical Law", *Foundations of Physics*, **3**, 139–168, (1973).
6. K. Pribram, "Toward a Holonomic . . .", *Op. Cit.*
7. J. Battista, "The Holistic Paradigm and General System Theory", *General Systems*, **XXII**, 65–71, (1977).
8. G. Bateson, *Steps to an Ecology of Mind*, (Ballantine Books, New York, 1972).
9. W. Harmon, "The New Copernican Revolution", In C. Muses and A. Young, (Eds.), *Consciousness and Reality*, (Discus, New York, 1974).
10. R. Lifton, "From Analysis to Form: Towards a Shift in Psychological Paradigm", *Salmagundi*, **28**, 43–78, (1975).
11. MacKay D., *Information, Mechanism and Meaning*, (M.I.T. Press, Cambridge, Mass., 1972).
12. A. Khinchin, *Mathematical Foundations of Information Theory*, (Dover, New York, 1957).
13. MacKay D., *Op. Cit.*

14. J. Battista, "The Science of Consciousness", In: K. Pope and J. Singer, (Eds.), *The Stream of Consciousness: Psychological Investigations into the Flow of Private Experience*, (Plenum Press, New York, 1978).
15. C. T. Tart, "States of Consciousness and State-Specific Science", *Science*, **176**, 1203–1210 (June 16, 1972).
16. S. Schacter and J. Singer, "Cognitive, Social and Physiological Determinants of Emotional State", *Psychological Review*, **69**, 379–399, (1962).
17. R. Anderson, "A Holographic Model of Transpersonal Consciousness", *Journal of Transpersonal Psychology*, **9**, 119–128, (1977).
18. I. Bentov, *Stalking the Wild Pendulum: On the Mechanics of Consciousness*, (E. P. Dutton, New York, 1977).
19. J. Banquet, "Spectral Analysis of the EEG in Meditation", *Electroencephalography and Clinical Neurophysiology*, **35**, 143–151, (1973).
20. E. Gellhorn and W. Kiely, "Mystical States of Consciousness: Neuropsychological and Clinical Aspects", *Journal of Nervous and Mental Disease*, **154**, 399–405, (1972).
21. C. Domash, "The Transcendental Meditation Technique and Quantum Physics", In: D. Orme-Johnson, (Ed.), *Scientific Research on Transcendental Meditation*, Vol. **1**, (Weggis, Switzerland, 1976).
22. J. Battista, "The Science of Consciousness", *Op. Cit.*
23. R. Sperry, "A Modified Concept of Consciousness", *Psychological Review*, **76**, 532–536, (1969).
24. R. Sperry, "Mental Phenomena as Causal Determinants in Brain Function", In: G. Globus, (Ed.), *Brain and Conscious Experience*, (Plenum Press, New York, 1976).
25. I. Bentov, *Op. Cit.*

11. DIALECTICAL PRAGMATISM: A PROGRESS REPORT ON AN INTERDISCIPLINARY PROGRAM OF RESEARCH ON DIALECTICAL INQUIRING SYSTEMS*

Ian I. Mitroff and Richard O. Mason

The authors show that dialectic reasoning can be construed as a concrete method of social science. It can be formulated as an operational methodology with widespread implications for all of the social sciences.

INTRODUCTION

This paper constitutes a progress report on an interdisciplinary research project into the nature of the dialectical inquirer. Over the past fifteen years the research of the authors has proceeded on two parallel but closely related fronts: the methodological and the philosophical. On the methodological side, there have been two central questions: (1) How can the dialectic as a philosophic concept be operationalized as a methodological tool of social science? (2) What does the dialectic both as a philosophic concept and as an operational methodology have to contribute to the theoretical understanding and practice of social science? On the philosophical side, the central question has been: what do the various attempts at operationalization contribute to our further understanding of the dialectic as a philosophic concept?

It should be acknowledged that the constant interplay between these two fronts is itself a partial expression of our own philosophical stance. As such, it derives from, among many, Ackoff[1,2], Churchman[3–5] and Dewey[6,7] with whom we are in deep sympathy. In a word, we do not believe that philosophy, or any other discipline for that matter, can proceed in isolation from the insights and knowledge of other disciplines or from contact with practice. The sphere of action is one of the significant arenas not only for testing our ideas but for their conceptual refinement as well. To twist an old saying, theory without practice is sterile and irrelevant; practice without theory is unreflective and hackneyed.

The paper essentially proceeds chronologically. It starts with the earliest attempts by ourselves and others to construe the dialectic as an operational methodology. In each case, we report the nature of the operationalization and its methodological/philosophical consequences. Since in most cases it is nearly impossible to separate the methodological consequences of each attempt from the philosophical (if the distinction is not artificial to begin with), we shall often report

*A fuller report of this effort can be found in *Creating a dialectical social science: Concepts, Methods and models*. (Reidel, Dordrecht, 1981).

them as one. The table of contents of our effort is: (1) Dialectic as Organization; (2) Dialectic as Communication Theory; (3) Dialectic as a Theory of Experimentation; (4) An Experiment on Training for Dialectic Thinking; (5) Dialectic as a Generalized Theory of Planning; (6) Dialectic as Argument; and (7) Dialectic as a General Theory of Social Science.

DIALECTIC AS ORGANIZATION

Within the last ten years, the field of organizational studies has undergone a significant revolution. The recognition has steadily grown that in a variety of ways complex organizations contain strong dialectical elements. One of the most important ways in which this is true is that complex organizations are composed of strongly contending factions. These factions often hold sharply antithetical conceptions of the internal organization and its relation to the external environment. As a result of differing world views, influential pockets of expert opinion develop. These groups give opposing interpretations of nearly every aspect of the organization.

If an organization can learn to take advantage of conflicting views, then the existence of such views can actually contribute to its long term survival and growth. The existence of sharply contending views can act as a much needed antidote to the natural tendency within most organizations to focus in and to secure allegiance to a single dominant point of view long after it has ceased to be appropriate or functional.

To the best of the authors knowledge, the first case study of the dialectic inquirer both as a naturally occurring phenomenon within an organization and as a tool for strategic planning was conducted by Mason.[9] The case involved a large equipment manufacturing and construction company. Two groups of equally powerful executives had diametrically opposing plans for the future development of their business. As a result of the basic conflict in plans, the groups were at an impasse. Neither group was able through rational argument alone to persuade the other of the reasonableness, let alone the 'truth', of its plan. Because both groups were roughly equal in power and status, neither was able to force its view through over the other. More frustrating still, data or facts by themselves were not able to verify or refute key claims in either argument. The reason is not that either case was impervious to empirical test, but rather because of the complexity, openness, and fluidity of real world arguments, each side was able to take almost any piece of evidence relevant to the issue and interpret it in its favour. Through intensive study of both groups, Mason was able to work backwards from the surface of the conflict (i.e., which of the two policies was best) to a deeper set of underlying issues. It turned out that these issues were the real source of the debate. Unbeknownst to themselves, both groups were making a host of assumptions about their internal management capabilities, the external market structure, and the general external social and political environment. While of critical importance, these assumptions were rarely articulated, let alone debated.

Mason discovered that it was possible (1) to state the assumptions of both groups in a dialectic format, (2) that this kind of format was pertinent to the kinds of complex issues managers and policymakers faced, and (3) that it proved helpful in understanding the complexity of the issue and deciding upon its ultimate disposition. A dialectic format meant that for every assumption which underlay the policy of one group, it was possible to construct a contrary or strongly antithetical assumption which underlay the policy of the other group. In this way, Mason constructed a dialectic between the underlying world views of both groups.

Next, a joint meeting was arranged and the dialectic was presented to both groups simultaneously. The underlying assumptions of the policies of both sides were arranged directly opposite one another on a large display chart. The meaning of each assumption was then illustrated by applying it to a common bank of data items, that is, a set of data which both sides had previously agreed was relevant in some not fully understood sense to both of their policies. The purpose of this step was to illustrate the fundamental point that, standing by itself, the data, that supposedly bore on the dispute, had no intrinsic, let alone singular, meaning. The data only took on meaning by being coupled to a world-view—in this discussion, a strongly related set of assumptions. Another way to put this is to say that meaning and interpretation are not separable. The data never settled anything in the previous disputes because it merely provided fuel for the underlying fire, the assumptions. This point is critical. The role of data takes on an entirely different role in complex, open-ended, fluid issues. Unlike issues that can be stated in terms of a single unambiguous, closed system, the purpose of data is not necessarily to say which of two competing sets of assumptions or hypotheses is 'correct'. Rather, its purpose is to surface the assumptions which underlie policies, to illustrate the role that assumptions play in the interpretation of data and their impact on resultant policies.

In summary, Mason's initial case study demonstrated the following: (1) a dialectic naturally arose in the context of one man's 'seemingly' more mundane activities, strategic planning within a large private organization; (2) the dialectic was a potentially fruitful tool for decision-making; (3) it shed important light on how to structure complex issues in dialectic format; (4) it suggested an initial set of categories or sub-issues around which a dialectic formed; (5) it shed light on how to conduct a dialectic as an exercise in small group process; and (6) it began confrontation of the complex issue concerning the appropriateness of a dialectic as a policy/planning tool versus other methods. The initial case study also confirmed an important earlier finding of the eminent social psychologist, Solomon Asch. In studying how people formed impressions of personality, Asch made the fascinating discovery that certain adjectives were much more potent than others. Thus, for instance, if Asch gave two people lists of ten or so adjectives which were identical in every respect save one, for example, one list contained the word 'cold' and the other contained 'warm', then the images of the persons represented by the lists would be perceived in very different terms. Mason found a similar result. In order for a dialectic to result, it was not necessary that the two (or more) plans or policies differ in every respect, but only in regard to significant issues. In this sense at least, the notion of incommensurability is *in part* a social psychological one.[10] Two world views can share an enormous amount (and thus make contact at significant points) and yet still be perceived or regarded by a group or person as antithetical.

Finally, in closing this section it should be emphasized that we have not been talking about the field of business policy exclusively. Writing in the field of clinical psychology, Levine[11] has come to many of the same insights and conclusions. The field of clinical psychology is due to the nature of the complex material it treats, prone to dialectic disputes between proponents of radically differing views of man, the nature of treatment, and so forth. In Levine's judgement the issues are too important and complex to be dealt with in the simplistic terms of one theory being right and another being wrong. Instead, Levine has proposed a dialectic airing of issues in much the same way proposed here.[11]

DIALECTIC AS COMMUNICATION THEORY

The case study described in the last section provided the basis for the authors' first attempt[12] to formulate more precisely and systematically the properties of a dialectical inquiring system (DIS). The heart of the attempt involved the use of Ackoff's concept of communication theory[1] and Bayesian probability theory.

The following concepts and ideas were at the basis of the representation. (1) There are three parties to the DIS, X, Y and Z. (2) X and Y represent contrary or antithetical policy experts who are charged with advising a decision-maker Z. (3) The world-views of X and Y may be characterized as the conjunction of a series of n atomic propositions; $W_X = \cap_{i=1}^{n} w_i$ and $W_Y = \cap_{i=1}^{n} \bar{w}_i$, where $\bar{w}_i$ = contrary of w_i. (4) There exists a common data set D such that when w_i and $\bar{w}_i$ respectively operate on it, opposing interpretations or conclusions c_i and $\bar{c}_i$ result. (5) X and Y are deadly enemies. This feature is represented in the model by the following conditions: (a) for X, $p(w_i/c_i) = p(w_i/\bar{c}_i) = 1$ and $p(\bar{w}_i/c_i) = p(\bar{w}_i/\bar{c}_i) = 0$; and (b) for Y, $p(w_i/c_i) = p(w_i/\bar{c}_i) = 0$ and $p(\bar{w}_i/c_i) = p(\bar{w}_i/\bar{c}_i) = 1$. In words, nothing X and Y say (c_i and $\bar{c}_i$ respectively) is enough to convince their opponent of the truth of their position. In Bayesian terms, the likelihood ratio is zero for one's deadly enemy. As a result, the amount of information that passes back and fourth between X and Y is zero. (6) The purpose of the debate is not for X and Y who are rigidly committed to their positions but for Z who is initially defined to be neither X nor Y. (7) As a result of witnessing the implications (c_i and $\bar{c}_i$) X's and Y's positions, Z comes to form his own position or world-view W_z that is an elementary synthesis of the positions of X and Y. For instance, for $n = 3$, one such elementary synthesis is: $w_1 \cap \bar{w}_2 \cap \bar{w}_3$. The probability of any particular elementary synthesis is computed through a complex application of Bayes' Theorem. In effect, the probability of any particular synthesis is a function of Z's likelihood ratios or the relative influence of X and Y on Z. (8) Given the relative influence of X and Y on Z, it is possible to study the amount of information that is communicated to Z as the result of the dialectic debate between X and Y. (9) The model shows that no matter what the initial likelihood ratios, the greatest amount of information transfer occurs for a Z who is originally indifferent between the positions of X and Y (i.e. $p(w_i) = p(\bar{w}_i) = 0.5$ but who later comes to agree completely with either X or Y. The model identified some important behavioral variables which influence a decision-maker's responses to a dialectic debate. As a result, it permitted comparison with the empirical evidence then available at the time.

DIALECTIC AS A THEORY OF EXPERIMENTATION

The formulation of the DIS in communication theory terms was appropriate for a first attempt because presumably one of the main purposes of a dialectic debate is to transmit information. However, for the design of behavioral science experiments, it turns out that a paradigm developed by the influencial psychologist Egon Brunswick[13] is more appropriate. The paradigm is known as the Brunswik lens model. In this model, a subject does not directly see a distal variable Y_D whose behaviour is to be estimated; rather, a subject sees a set of immediate cues X_i which are potentially correlated between themselves $r_{X_iX_j}$ and with $Y_D(r_{Y_D,X_i})$. The subject's task is to formulate and estimate Y_S of Y_D that is based on what is observed, i.e. X_i, $i = 1$ to n. The correlations $r_{X_iX_j}$ and r_{Y_D,X_i} can be set or fixed by the experimenter and hence the model Y_S that the subject uses to predict or estimate Y_D can be studied.

In the earliest Brunswik lens models, both Y_D and Y_S were considered to be multiple linear regression models of the X_i. Given two different multiple linear regression models, it is relatively easy to derive a formula for R_{Y_D,Y_S}, the criterion variable or performance of the subject S in estimating the distal variable Y_D.

The Brunswik lens model was applied recursively to develop a more complex model of DIS. First, two antithetical policy experts were defined in terms of a dialectical weighting of the common cues X_i. That is, in the lens model the common data set to which both experts address themselves is X_i. The experts then apply a very different weighting to the common data to derive different estimates of Y_D. It turns out that the simplest defining condition for a pair of dialectical policy experts or advisors to S is the formula $\Sigma_i \beta_i^1 \beta_i^2 = 0$, where β_i^1 are the regression weights of the first policy expert, etc. This condition says that the weights of the two experts are orthogonal to one another.

Second, the cues that the subject sees are no longer the X_i directly but Y_1 and Y_2, the judgements or models of the two experts applied to the so-called 'raw' or 'primitive' data X_i. The point is that the 'cues' a decision-maker sees in everyday life are vastly more complex than those which are seen in the typical psychology experiment. In a DIS what a decision-maker 'sees' are the complex judgements of opposing policy-experts. The question is how a decision-maker comes to form his own model of Y_D given the conflicting advice of Y_1 and Y_2.

The results of the exercise show that it is possible to formulate a model of a subject's behaviour Y_S that is a multiple linear regression of the models of the two conflicting experts, Y_1 and Y_2. Further, Y_S can be identified with the synthetic behaviour of S. The exercise further shows that correlation between the cues $r_{X_iX_j}$ and the condition set definite restrictions on Y_S, i.e. on the simple synthetic policies of *S*. This result while not surprising is important nonetheless. It says that baseline models of the simplest form of synthesis can be formulated and the behaviour of such models can be studied.

Before leaving this section, it is important to emphasize that Kenneth Hammond and his colleagues[14] have performed extensive studies of the Brunswik lens model in both the laboratory and real-world policy settings. Indeed, Hammond's work served as a powerful stimulus for our own work. Hammond's work shows clearly that the Brunswik lens model is a powerful paradigm for the study of complex human behaviour in a conflict mode.

TRAINING FOR DIALECTIC THINKING

Attempts at applying the Brunswik lens model of the DIS in experimental settings soon revealed a basic difficulty which penetrated to the heart of the DIS. The idea of dialectic thinking, that a concept and its antithesis can both be highly *plausible*,[15] is foreign to most people, in our first culture at least. This and two other highly related facts made the study of how subjects react to the DIS difficult to carry out.

Social scientists have repeatedly found that the topic of conflict is a highly complex phenomenon in its own right, and further, that people differ markedly but systematically in their styles of reacting to and handling conflict.[16] Some subjects could be expected on the basis of previous findings to be receptive to the basic idea of the DIS. Since this was the case, the responses of these subjects to the DIS could willingly be studied. Other subjects resisted the whole idea of conflict as a basic epistemic mechanism. The reactions of these subjects to experiments on the DIS were largely negative. As a result, the authors turned their attention to a more fundamental issue: Could subjects in general, no matter what their initial personality disposition towards conflict, be trained to understand and to appreciate the concept of the DIS and dialectic thinking?

The results of the investigations to the preceding question have been reported in detail elsewhere.[17] In general, the answer is a strong 'Yes'. Subjects *can* be trained both to understand and to appreciate better the concept of the DIS. The results are all the more powerful in that the training sessions were conducted on a computer. That is, a subject's attitudes towards dialectic thinking were assessed both before and after exposure to a training session programmed on a computer. Equally powerful is the finding that the results were short-lived. The training sessions were extremely powerful in effecting or shifting a strong predisposition towards dialectic thinking; however, the decline in the belief of the newly acquired thinking was equally powerful.

The message is that subjects can be taught to appreciate that on complex issues they are wise to listen to the stories of competing experts, if only for the reason that this is extremely helpful in better understanding the assumptions which underlie the positions of experts. Indeed, this is one of the fundamental premises of the DIS, that a dialectical presentation of issues is critical to a richer understanding of the assumptions upon which they rest. The message, however, is also that subjects need continual reinforcement if they are to assimilate this different way of thinking.

DIALECTIC AS A GENERALIZED THEORY OF PLANNING

As all inquiries do, the initial case study conducted by Mason raised as many subsequent questions as it provided first responses. Among the many questions which flowed were these: Where does a dialectic come from? Where do the issues come from that form a dialectic? Are there only *two* contending positions? If the differences in viewpoints are not explicit but underlying, how can one shape them into a dialectic and if need be, create a dialectic?

Over the years that we experimented with the DIS, the most dramatic breakthrough with regard to the above issues occurred in the area of small group processes. The breakthrough came through the discovery that how a group is structured greatly affects, if not almost determines, its behaviour. For instance, there exist a number of instruments which purport to measure an individual's personality and/or cognitive style, and in fact, the reliability and validity of a number of them is quite good for various predictive and counseling purposes.

In particular, the authors have worked with a specific personality typology as measured by the Myers-Briggs instrument.[18] We have found that placing all those individuals of the same personality 'type' into a single and distinct group exerts a very dramatic effect upon the behaviour of the resultant small group that is formed. Because all the individuals in a particular group tend to see the world in similar terms, the outlook of the group as a whole is generally intensified even more over the outlooks of the individuals. For the very same reason, interpersonal conflicts between individuals are generally reduced so that they can concentrate their energies almost exclusively on the cognitive or intellectual task at hand. In most groups in everyday life, this is not the case. The groups are so divided in outlook that the majority of energy has to be expended in working through personal and emotional differences before they can ever get to the task. Again, because of the similarities in outlook, the blind spots of the groups also tend to be intensified. As a result, what each group considers as a natural 'given', their opposite personality type group considers an 'unwarranted taken'.

It soon became apparent that this simple sorting device accomplished a number of important aims.

First, it allowed the participants as a whole to take advantage of both their similarities and their differences. One of the most persistent findings in the social science literature is that the effectiveness of small group problem solving is a function of how a group manages or takes advantage of its similarities and its differences. The sorting of individuals into separate groups is a concerted way of making use of both heterogeneity and homogeneity. The homogeneity *within* a group allows it to function effectively without friction and strife. The heterogeneity *between* groups is an effective way of taking advantage of the differences in thinking among the participants as a whole. It allows the groups to debate one another but only after they have first had the opportunity to develop their own position without having to attend to the diverting challenges of their opponents.

Secondly, the sorting device allowed a large number of individuals with different views drawn from throughout an organization to participate in the planning/decision-making process. As problems have become increasingly more complex in their definition, let alone solution, widespread participation is no longer a luxury. It is necessity. The organizational behaviour literature has shown repeatedly that unless individuals are meaningfully involved in the definition of a problem, they will not give their whole-hearted energies in implementing a solution, especially one that is thrust upon them. The group sorting technique allows a large and diverse group of individuals to break itself down into a small number of manageable subunits.

The next critical breakthrough came with the discovery of a simple *behavioural* process for surfacing the assumptions that underlie a group's position or outlook. In brief, the assumptions that underlie a group's position policy, or definition of a problem/issue are the properties that must be posited about the critical actors that comprise a complex social system. These actors are known as the *stake*holders. The stakeholders are the collection of critical parties who significantly affect and in turn who are affected by any policy. It is beyond the limits of this paper to describe in detail how significant stakeholders are identified and how the presumed properties pertaining to them are surfaced and then treated. Even further beyond the scope of the paper is how the difference in group perceptions (stakeholders, assumptions, policies) can be *behaviourally* synthesized into a coherent policy. The point is however, that the synthesis in world view is not necessarily a logical process but a social-psychological one.

Finally, it should be noted that there is not a single way of forming groups to engage in a dialectic.[10] People can be sorted on the basis of their personality, their interest in certain tasks or issues, or both. Which method is used depends upon the problem and the organization. As a general rule, we have found that no matter which method is used, four distinct groupings always tend to appear. One group tends to take a distinct short-term, operational, technical view of issues, another, a long-term, holistic, technical standpoint. A third group tends to assume a long-term, holistic behavioural and/or humanistic view; a fourth, a short-term, intensely personal, humanistic view. These groups seem capable of taking any issue—no matter what—and defining it in very different terms. The different groups also embody one of the prime features of a dialectic. The policy of each group is defined by what it affirms as well as by what it negates or what it opposes. Thus, the full meaning of the policy of any one group cannot be understood solely by reference to that group but must be accomplished by the additional reference or comparison with its opposition.

We had achieved one of our long sought-after aims, the third group tends to assume a long-term, holistic behavioural and/or humanistic view; a fourth, a short-term, intensely personal humanistic view. These groups seem capable of taking any issue—no matter what—and defining it in very different terms. The different groups also embody one of the prime features of a dialectic. The policy of each group is defined by what it affirms as well as by what it negates or what it opposes. Thus, the full meaning of the policy of any one group cannot be understood solely by reference to that group but must be accomplished by the additional reference or comparison with its opposition.

We had achieved one of our long sought-after aims, the construction of a simple *behavioural* method for structuring a dialectic debate. We have repeatedly emphasized the word 'behavioural'. For a method to work, it must be capable of specifying in precise detail what people can do, not to mention how and when, if those who do not understand the nature of dialectic reasoning are nonetheless to be able to use it as a social problem solving device.[19]

DIALECTIC AS ARGUMENT

Up to this point, we have deliberately avoided any references to the notion of a dialectic logic. The essential reason is, as is well known, that the whole subject is fraught with innumerable difficulties. This area also represents the most recent of our attempts. As such, we can only outline briefly here the nature of the effort.

Given our concept of the DIS, the notion of a dialectic logic cannot be developed independently of a larger framework of argumentation. The basic reason is that the dialectic is a *process* of reasoning that unfolds over time. As such, the path (the considerations) by which one has reached a particular conclusion is as important as the conclusion itself. Second, it is within this larger framework that the concept of the *plausibility* of an assertion and its dialectical opposite or negation needs to be defined.

Our effort has proceeded by attempting to join the work of two eminent philosophers, Stephen Toulmin and Nicholas Rescher. In an earlier work,[20] Toulmin proposed an interesting framework in terms of which to capture the dynamics of reasoning as it actually occurs within the various processions. In more recent work,[15] Rescher has defined the concept of plausibility and shown how it can be used to handle sets of propositions containing strong inconsistencies. We have not only attempted to join these two concepts but to create still a more precise rendering of the DIS.

In Toulmin's framework, every argument terminates in a Claim. The Claim is thus the conclusion of an argument and is only as good as the structure which supports it. The Evidence is that part of the argument which represents all the data, facts, or grounds in support of the Claim. The part which justifies the movement or transition between the Evidence and the Claim is termed the Warrant. If the Warrant is not accepted at its face value, and it often is not, then underlying it is the Backing. In effect, the Backing is the deeper set of background reasons for believing in the Warrant.

One of the most critical elements in the framework is the Rebuttal. The Rebuttal is a single element which more than any makes the possibility of a dialectic debate a feature of *every* argument. The Rebuttal is the series of challenges to any part of the main argument.

It is a simple matter to show that if for simplicity one limits oneself to the first order predicate calculus, the different forms of the Rebuttal may easily be tabulated. For instance, one of the most prevalent forms of the DIS is the case where two parties both agree that a certain data set is relevant to their deliberations. However, one group asserts that C follows from D (hence, their Warrant is $D \rightarrow C$) while the other asserts that $\sim C$ follows from the very same D (hence, their Warrant is $D \rightarrow \sim C$). Depending upon which argument is looking at $D \rightarrow C$ or $D \rightarrow \sim C$ is the Rebuttal.

The machinery of Rescher comes into play in the assignment of the relative credibilities of plausibilities (*pl*) of $D \rightarrow C$ versus $D \rightarrow \sim C$. In the broader sense, Rescher introduces the concepts which allow one to sort a larger, single, *in*consistent set of propositions into a number of smaller consistent sets of propositions. These are maximally consistent sub-sets. In the simple example above, there are two such sets: $\{D, D \rightarrow C\}$ and $\{D, D \rightarrow \sim C\}$. Rescher also introduces various rules, which are analogous to game-theoretic rules, for deriving $pl(D \;\&\; C)$ from $pl(D)$ and $pl(D \rightarrow C)$.

One straightforward interpretation of the DIS is this: Two positions are dialectical adversaries if (a) they represent diametrically opposed conclusions, e.g., $D \;\&\; C$ and $D \;\&\; \sim C$, and in addition, (b) both are equally high in their credibility; i.e., $pl(D \;\&\; C) = pl(D \;\&\; \sim C)$.

Efforts are currently underway to extend the concepts even further by assigning a meaning to $pl(C \;\&\; \sim C)$ such that $pl(C \;\&\; \sim C)$ is not low. This occurs if it can be given a meaning in some new argument structure corresponding to a synthesis of frameworks or earlier arguments.[21]

DIALECTIC AS A GENERAL THEORY OF SOCIAL SCIENCE

It is beyond the scope of a limited discussion to show in detail how the arguments advanced throughout this paper can be assembled to formulate a dialectical theory of social science.[10] In brief, it can no longer be denied that all of science, but especially the social sciences, are racked by intense disputes over the nature of what constitutes proper methodology, the 'essence' of man, the role of the social scientist, etc. These disputes are not amenable to easy settlement for they proceed from very different and deeply held background assumptions. It can be shown that the conflicts between different traditions can be 'tentatively' understood in terms of an on-going dialectic between at least four very different conceptions of social science.[10]

In the spirit of a dialectic, there is no 'final', in the sense of 'fixed', ending to an effort of this kind. It is continually on-going. If there is a firm conclusion to which we hold, it is this: one can always argue endlessly why philosophy and social science are separable activities, but we do not believe that such separation is wise. We do not believe that the DIS can be understood by philosophers or social scientists acting alone—even though the debate between them is always to be encouraged.

REFERENCES

1. R. L. Ackoff, 'Towards a behavioural theory of communication', *Management Science* **4**, 218–234 (1958).

2. R. L. Ackoff and F. Emery, *On Purposeful Systems* (Aldine-Atherton, Chicago, 1974).
3. C. W. Churchman, *The Design of Inquiring Systems* (Basic Books, New York, 1979).
4. C. W. Churchman, *The Systems Approach and its Enemies* (Basic Books, New York, 1979).
5. C. W. Churchman, *Theory of Experimental Inference* (Macmillan, New York, 1948).
6. J. Dewey, *Experience and Nature*, (Open Court, La Salle, Illinois, 1925).
7. J. Dewey, *The Quest for Certainty* (Macmillan, New York, 1929).
8. K. R. Hammond *et al.*, 'Probabilistic functioning and the clinical method', *Psychological Review* **62**, 255–262 (1955).
9. R. O. Mason, 'A dialectical approach to strategic planning', *Management Science* **15**, B-403-B-414 (1969).
10. I. I. Mitroff and R. H. Kilmann, *Methodological Approaches to Social Science* (Jossey-Bass, San Francisco, 1978).
11. M. Levine, 'Scientific method and the adversary model', *Amer. Psychologist* **29**, 661–677, September 1974).
12. I. I. Mitroff, 'A communications model of dialectical inquiring systems', *Management Science* **17**, B-634-B-648 (1971).
13. I. I. Mitroff, 'A Brunswik Lens model of dialectical information systems', *Theory and Decision* **5**, 45–68 (1974).
14. N. R. Hanson, *Perception and Discovery* (Freeman, Cooper, San Francisco, 1969).
15. N. Rescher and R. Manor, 'On inference from inconsistent premises', *Theory and Decision* **1**, 179–217 (1970).
16. R. H. Kilmann and K. W. Thomas, 'Four perspectives on conflict', Working Paper No. 86 (Graduate School of Business, Pittsburgh, PA, University of Pittsburgh, 1974).
17. I. I. Mitroff and J. Nelson, 'An experiment on dialectic information systems', *J. Amer. Soc. Inf. Science* **25**, 252–262 (1974).
18. I. B. Myers and K. G. Briggs, *Myers-Briggs Type Indicator* (Palo Alto, California, Consulting Psychologists Press, 1977).
19. I. I. Mitroff and J. R. Emshoff, 'On strategic assumption making: A dialectical approach to policy and planning', *Academy of Management Review* **4**, 1–12 (1979).
20. S. Toulmin, *The Uses of Argument* (University Press, Cambridge, England, 1958).
21. I. I. Mitroff and R. O. Mason, *Creating a Dialectical Social Science: Concepts, Methods, and Models*, (Dordrecht, Reidel, 1981).

12. TELEOGENIC SYSTEMS THEORY

Norman A. Coulter, Jr. and Albert L. Johnson

SUMMARY

Teleogenic systems theory, the theory of systems which can create their own goals, provide a basis for a metasystem methodology which may be incoporated in the metasystem paradigm. The change agent, who is both observer, actor and participant, functions in terms of the multiple actualities of complex human systems. He enables and facilitates changes aimed at promoting *synergy* (any property of the system that promotes two or more goals, functions, or values, while impeding none). The teleogenic system theory methodology may be useful in conflict resolution and in creating situations which predispose to co-operative processes in groups and organizations where diverse actualities prevail.

INTRODUCTION

Teleogenic systems theory deals with systems that are capable of generating their own goals. Such systems differ from engineering control systems whose goals are fed in by outside operators, and from power hierarchies whose goals are set at top hierarchical levels within the system. They have many interesting properties, some of which are described elsewhere.[1-6]

In developing teleogenic systems theory, it was found that the role of the observer was critically important.[7] An observer whose perspective excludes teleological processes will be unable to perceive goals where they exist. Fortunately, cybernetics and management science have made the study of such processes scientifically acceptable again.[8] We found that it was useful to distinguish three types of teleological systems, each more advanced system including the more primitive ones as special cases. These are:

1. Teleonomic systems, which are capable of goal-seeking behaviour, but whose goals are fed into the system from outside.
2. Teleozetic systems, which are not only teleonomic, but which also can choose their own goals from a repertoire with which they were initially endowed.
3. Teleogenic systems, which are not only teleozetic, but which also can create new goals.

Here, we wish to explore the possible use of teleogenic system theory in the development of a metasystem methodology. As will be shown, the purposes of the observer are important factors in determining what he observes; and when the observer is also a designer/engineer/planner, his purposes, and the context from which those purposes are derived, also determine what he does. Hence, teleogenic systems theory is relevant to the metasystem paradigm. We will first consider the context in which teleogenic systems theory is embedded by discussing actuality theory. The role of the observer who is also an actor and participant in the metasystem paradigm is then described. Finally, it is proposed that the promotion of *synergy* be the global goal of the metasystem change agent.

ACTUALITY THEORY

To paraphrase Voltaire: If there were no reality, it would be necessary to invent one. The human mind, designed by evolution, is organized to promote survival in an environment. Its basic structure requires for there to be objects 'out there' that have an autonomous existence, independent of the mind that perceives them. Even Berkeleyan idealism postulates the existence of objects in the 'mind of God', independent of human awareness.

Yet the world we live in, with tangible objects and real people, is actually a projection of our brains. It is like three-dimensional television with holograms and stereophonic sound. We are 'certain' it is 'real' because the projection is remarkably accurate; the information processing networks of the brain continuously test for accuracy and consistency and instantly correct errors.

This is demonstrated by optical illusions and other phenomena of visual perception.[9] What is true for vision is equally true for other perceptual systems. On the basis of sense data, the brain *posits an entity* to exist in its 'actual World,' a 'working hypothesis' that the brain rapidly and continuously tests in ways we do not yet fully understand. As long as no errors or inconsistencies are found, the brain accepts the entity as 'real' or 'actual'. Subjectively we experience a conviction that the object is *really there.*

Let us call a posited entity that has been accepted as 'real', 'true', 'right', or in some other sense 'valid', a *referent.* An actuality, then, may be defined as a set of referents. The term 'actuality' is used rather than 'reality', because an actuality may include error, illusion, delusion, hallucination, or social convention. It is evident that there are many actualities. Each individual has a personal actuality. If two individuals communicate or otherwise interact, their personal actualities will not in general totally coincide, The 'area of agreement' of their personal actualities constitutes a 'consensus' or 'social' actuality. New referents, sanctioned by agreement, may emerge or be created in the course of their communication and interaction. These new referents then become part of the evolving personal and social actualities of the individuals involved.

To be part of an actuality, a referent must have three properties:

1. A *characteristic* or *membership* function, which affirms the 'validity' of the referent with respect to the actuality. For a personal actuality, this function may be membership in a perceptual category, successful completion of one or more 'validity checks', etc. For a social actuality, this function may be tacit or formal agreement by all parties sharing the actuality.
2. A *discriminant,* which enables the referent to be distinguished from other referents of the actuality. If a sound is heard and an object is seen, they are perceived as two distinct referents, unless the seen object is cognized as the source of sound. Their discriminant functions being the same, they are then perceived as the same referent.
3. *Autonomous existence,* independent of the conscious awareness of the individual or group. The entities of dreams or fantasies are not ordinarily regarded as referents in an actuality.

Thus far, we have considered the referents of an actuality as a collection of distinguishable elements. However, various relationships among these referents exist. We may regard a *system* as a set of referents in an actuality, organized according to one or more ordering relations, and relatively separate or separable from other referents of the actuality.

Thus, for example, the solar system consists of *referents* (sun, planets, their satellites, asteroids, comets, etc.) *ordering relations* (positions, velocities, orbits, the law of gravitation, etc.); and separation from other celestial objects beyond the influence of the gravitational field of the sun and outer planets. The solar system is embedded in an actuality — the physical universe.

The perspective of the observer is tacitly involved in the definition of a system in an actuality. Thus each referent of the solar system, viewed from a smaller scale perspective, is itself a system. From a different perspective, each referent is also a subsystem of the solar system. In turn, the solar system is a subsystem of the Milky Way galaxy.

Thus in at least two ways the role of the observer has a subtle and tacit influence on an actuality: (1) the autonomous existence of referents — *esse est posse non percipi* — and (2) the perspective of the observer in defining the system to be observed. In what other ways is the role of the observer important?

ROLE OF THE OBSERVER/ACTOR/PARTICIPANT

First, let us consider the context in which the observer observes. What are his purposes? It is essential at the outset to correct past tendencies to consider only one of the observer's purpose while all others are treated, if at all, as aberrant error factors to be controlled. For instance, it has been said that the purpose of the observer, acting as a scientist, is to describe and explain the observed phenomenon. Descriptions should be based upon minimal interference with the 'thing' being observed, and measurements of its characteristics and/or behaviour should be as accurate as possible. Explanations, however, involve the 'explainer' in many levels of interference with the 'thing' to be explained. He not only hypothesizes how the thing will act under specified conditions, he also manipulates it to test for the correctness of his guess. In so doing the observer *of* the 'thing' has become an actor *on* the 'thing'.

Scientists, like all other observers, are never completely isolated from their social support systems. The process of confirmation and correction among scientists is an integral part of the scientific method which probably contributes greatly to the rapid growth of knowledge by eliminating many false leads. Scientists are also dependent upon financial and other support mechanisms in spite of their often-cited 'freedom of inquiry' prerogatives. Some inquiries are starved while others are engulfed with affluence by those possessing control over support mechanisms. Quantification may, through publish-or-perish dicta, supported by citation indices, reward orthodoxy while penalizing originality.

Scientists are not immune to the use or abuse by others of the knowledge they create, as physicists and geneticists have so sorely learned. Even among their limited numbers, however, there is no overriding set of ethical standards or political views which would guarantee a beneficence for all people.

There are a number of tacit assumptions made by scientists that merit consideration. These include:

1. The system observed is not unique, but one of a number of identical systems.
2. The system is observable.
3. The system, during experiments, is controllable.

The assumption of non-uniqueness implies that any system that is unique, or that undergoes a sequence of transformations which contain unique components, cannot be completely understood by the scientific method described above. This does not mean that such systems cannot be understood, in some sense; it only means that the scientific method can at best give only a partial understanding. Possible examples of such systems are: the history of a nation, race, class or mankind as a whole; the life-experience of a human mind; the evolution of biological systems; individual organisms of a species (whose genetic endowment is in general unique).

Unfortunately, the implications of this tacit assumption are often overlooked. This leads to the view that what is unique does not exist, or is unimportant.

A comparable problem results from the assumption of observability and controllability. It is well known that multivariable systems may include components that are uncontrollable or unobservable or both.[10] It has also been shown that teleogenic systems necessarily include at least one uncontrollable component.[2,11] An example of a system with uncontrollable and unobservable components is shown in Figure 1.

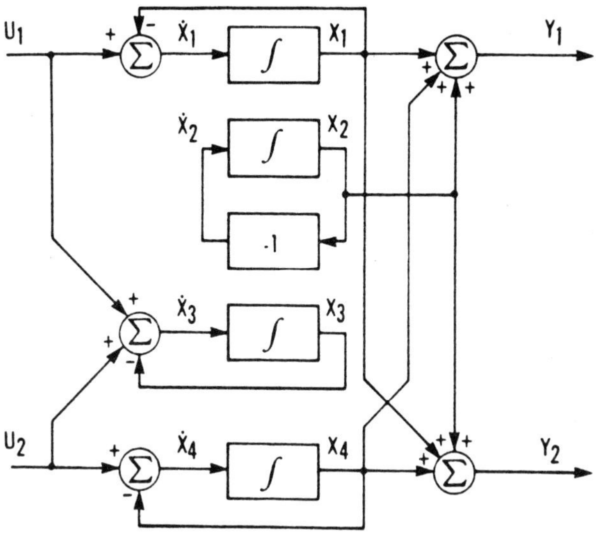

Fig. 1. State variable X_2 is uncontrollable – independent of inputs U_1 and U_2. State variable X_3 is unobservable – without effect on outputs Y_1 and Y_2.

The assumptions of complete observability and controllability imply that any system containing an unobservable or uncontrollable state variable cannot be completely understood by the scientific method described above. Since human beings, groups, and other social entities are teleogenic, they cannot be completely controlled. That they also cannot be completely observed is a primary datum of experience.

Because teleogenic systems generate their own goals, they *participate* in communications and interactions of a network in ways that are overlooked in organization charts and mechanistic models that ignore the teleogenic quality of their components.

There is also an extremely important moral, as well as practical, issue involved when one or more persons or systems attempt, even though it is not possible, to control other persons or systems. The Nuremburg trials sensitized many people to the dangers of such attempts, and Milgram's *Obedience to Authority*[12] and Yablonsky's *Robopaths*[13] continue to remind us how prevalent such efforts are, even in a country not at war.

In the foregoing, the role of the scientist as observer/actor/participant has been described, and the limitations implied by the tacit assumptions of the scientific method discussed. The purpose of the scientist is to describe and explain. What about other observers whose purpose is different?

Consider the designer or engineer. His purpose is to design a system to perform a specified function. He seeks to prescribe a system and to actualize it. In doing so, he uses knowledge generated by scientists, together with design methodologies and practical knowledge gained from experience. Characteristically, he operates according to a goal hierarchy. He has an overall or global goal: to design a system to perform a given function. The system itself has a goal — its function, which may be described as a set of actions that fulfill a goal. A watch is designed to determine time and to display it, for example. The global goal of the designer-engineer generates a hierarchy of subgoals aimed at solving problems related to the design of the system.

The engineer-designer is also a participant in a larger social system, subject to constraints analogous to those faced by the scientist. The global goal is usually assigned to him; if he himself creates a global goal, he is constrained by his status in the organization he works for to submit 'ownership' of the idea to the firm. The ultimate destiny of the system he designs is also largely outside his control.

The organizer-manager, as often conceived, has a different purpose. The system he deals with is 'soft' and extremely complex. It involves human beings and groups as well as materials and systems designed by engineers. It also includes laws and regulations, communications and channels of command, economic and financial processes, etc. He occupies a position in a power hierarchy, responsible to those above, in command of those below. In addition, he is involved in competitive and co-operative relationships with others outside his chain of command (as well as within it).

The organizer-manager observes, acts, and participates; but unlike the scientist or engineer-designer his observations are selective, his actions guided more by heuristics than by algorithms, and his participation characterized by a stream of decisions based on new data received from other parties and his rapid, intuitive evaluations of the impact his communications and actions have on those parties. Novel events continuously emerge; and while attention to particular details is often required, he must constantly view his system holistically, keeping in mind a constantly changing 'big picture'. While this concept of the organizer-manager is widespread, its actualization usually leads to 'irrational' conflicts and other social pathologies.

One of the factors perpetuating these social pathologies is the concept of a designer as a creator of systems with one specified function. At the teleonomic system level this concept may be tolerable just so long as any of the parts of the system are not humans. At the teleozetic system level, as so well documented by Gross,[14] there is always a hierarchy of purposes to be attained rather than a single-minded unitary function to be performed. This multiplicity of purposes is developed by Papenek[15] as he attempts to *Design for the Real World,* and as though humans really do matter. At the teleogenic system level, we are just beginning to reconceptualize the design concept so as to free us from superimposed controls. Burrell and Morgan[16] among others, challenge the functionalists' domination of our thoughts about social systems; Burns[17] turns *leadership* into a co-operative versus directing process; and Pfeffer,[18] Illich,[19] and Schumacher,[20] all help us to see how continuously we redesign the systems we participate in rather than serve purely as subjugated system parts.

One feature of these different types of system observer stands out: the primary role of purpose. Different global purposes distinguish these types from each other; and purposes continually guide the observations, actions, and participations performed. This suggests that the role of purpose is central to the metasystem paradigm.

THE ROLE OF PURPOSE IN COMPLEX HUMAN SYSTEMS

To a mechanistic scientist, purpose does not exist; or, if it does, it is irrelevant, an anachronistic concept that belongs in the wastebasket of useless metaphysical ideas. The Aristotelian 'final cause', a future state retroactively intervening in the causal chain of events, does indeed seem paradoxical. And it must be acknowledged that remarkable advances in scientific knowledge have been achieved without explicit reference to teleological ideas.

The advent of cybernetics changed this. As noted by Wiener and others, the input to a servomechanism can be regarded as a goal signal, *representing* a future state of the output of the servomechanism. This is illustrated in the right part of Figure 2. A signal representing the current position of the output is fed back (opposite in sign) to the comparator, which compares it with the goal signal. If the two do not coincide, an 'error signal' from the comparator drives the servomechanism until the output signal

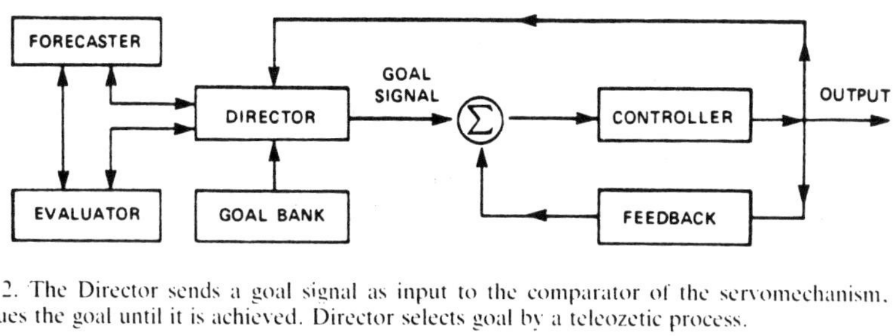

Fig. 2. The Director sends a goal signal as input to the comparator of the servomechanism, which pursues the goal until it is achieved. Director selects goal by a teleozetic process.

and goal signal coincide. When this happens, the error signal from the comparator is zero, and the servomechanism stops changing the output. The goal has been achieved. Moreover, if any extraneous force changes the output, the servomechanism will be activated again, resisting the change. Although human purposive behaviour is admittedly far more complex, the action of a human committed to a goal is essentially like this. Cybernetics thus has made the concept of purpose scientifically acceptable again. It is a *representation* of a future event, not the future event itself, that occupies a key position in the causal chain.

A servomechanism, it should be noted, is basically different from an ordinary machine. An ordinary machine is not *goal-seeking;* it simply converts energy into work. When a machine is used for a purpose, the comparator and feedback loop are located *within the human operator*. Nevertheless a servomechanism only accounts for a part of purposive behaviour. The goal is fed into the servomechanism from the outside. It then pursues the goal until it is achieved. In teleogenic system theory, a servomechanism is an example of a teleonomic system.

Where do the goals come from? To account for the selection of goals, and the generation of new ones, an additional component is needed. This is represented in the left part of Figure 2.

This illustrates a simple teleozetic system, endowed with a *set* of potential goals, stored in the Goal Bank. The Director selects one or more of these potential goals, and transmits them to a Forecaster, which forecasts the outcome if each potential goal is adopted. These outcomes are fed to the Evaluator, which assigns a worth value to each potential goal. The Director then selects the goal with highest worth value (assuming it is above 'decision threshold'). This goal is then transmitted as a *goal signal* to the comparator of the servomechanism which then pursues the goal.

A more elaborate system, capable not only of adopting a goal from a set of potential goals, but also generating new potential goals, has been described elsewhere.[3,4] Such a system is called a *teleogenic* system. Although far more complex than the system depicted here (which can be simulated on a computer), a human being clearly has teleogenic capability. This fact is of basic importance for a metasystem paradigm.

While the concept of 'goal' may be rejected by a mechanistic scientist, it is freely accepted by organizer-managers acting according to the power-hierarchical concept of social organization described above. Such a concept is basically teleonomic; and the organization is tacitly assumed to have a single global goal. But an organization consisting of a network of teleogenic components is clearly multipurposed; and in operation its structure is not fixed by organizational charts, but is continuously being modified and redesigned, not only by those at the top level of the power hierarchy, but at all levels. Inevitably, these various purposes come into conflict; and this leads us to the application of teleogenic system theory to the metasystem paradigm.

TELEOGENIC SYSTEM THEORY AS A METASYSTEM METHODOLOGY

We are now in a position to describe how teleogenic system theory may be used as a metasystem methodology. We shall follow the outline suggested by van Gigch.[21]

1. Description of field

A teleogenic system methodology applies to any system involving teleogenic components — such as human groups, organizations, or other social systems. It may be used by any party (individual or subgroup) whose purpose is to change the system in some way. The change may be small or large; the 'change agent' may occupy any status or fulfill any set of roles; the party may use existing channels of communications and interaction or create new ones.

2. Nature of domain.

The system of teleogenic components is regarded as being embedded in the multiple actualities of each component, the various common actualities among local subgroups, and the common actuality of the system as a whole. The actuality of the change agent is also considered.

3. Problems to which suited.

The methodology is applicable to problems of conflict resolution, to the partial or complete transformation of adversary mode situations into synergic mode situations, and to the initiation, growth, and development of evolutionary processes of the system which tend to enhance synergies among its various functional components and the system's global synergy.

4. Application of the methodology.

In addition to the ideas described above, there are several ideas that are basic to the application of the teleogenic system methodology.

a) *Synergy*. This is defined as any property of the system which promotes two or more goals, functions, values, etc. of the components, while impeding none. The global aim of the change agent is to promote synergy. The 'content' of the goals, etc. is not addressed by the methodology, although it is considered by the change agent in appraising actualities and potentialities. For example, if two goals clash, the change agent does not seek to 'take sides' or 'change goals', but to create and develop channels of communication and interaction which enable the parties each to achieve its goal without impeding the others.

b) *Dysergy*. This is defined as any property of the system which promotes one goal, function, value, etc. but impedes another. The term 'dysergy' is content free. It is used instead of 'conflict' because the latter term implies a struggle to be waged, an adversary to be overcome, etc. 'Dysergy', on the other hand, implies a problem to be solved, with due consideration for the interests and actualities of both parties. The reduction and elimination of dysergy facilitates, and is facilitated by, the promotion of synergy. They are opposite sides of the same coin.

c) *The Mode of Function Scale*. To aid in promoting synergy and reducing dysergy, a 'Mode of Function Scale' is used. This is based on appraisal of five modes of communication, action, and interaction that are distinguishable, by a trained observer, in the behaviour of individuals and groups. The modes are, from 'bottom' to 'top.':

i) *Identic*. This is the mode of a person under hypnosis. It is also the mode of a person or group driven by 'magical thinking'. The conditioning process that occurs in Pavlovian or operant conditioning is another manifestation of this mode of function.

The characteristic operation of the Identic Model is *identification.* Two mental contents are linked together as being *identically equal* — like the two sides of a trigonometric identity. There is no discrimination between the two contents — no matter how different they may be. There is no awareness of the linkage as such.

A witch doctor makes a voodoo doll of his enemy. He sticks a pin in the head of the doll — and his watching enemy instantly feels a piercing headache.

A bell is rung. Two seconds later, meat is placed in a dog's mouth. Saliva flows. This is repeated several times. Then, the bell is rung, but no meat is given. After a precise interval saliva flows. An identification has been established between the sound of the bell and the flow of saliva.

The Identic Mode is a primitive mode of function of the nervous system of vertebrates, including man. Many of the techniques of advertising and propaganda are designed to establish identifications in the minds of watchers or listeners. Knowing what to look for, an observer can soon detect Identic Mode operations in the communications and actions of other human beings. The parties observed are unaware of the linkage between the two contents, and do not discriminate between them.

ii) *Reactive.* This is the mode of an individual or group who is *driven* by strong emotion — fear, anger, hate, grief, lust, guilt, etc. The party is aware and makes simple distinctions. These are characteristically diametrically valued: good-evil, right-wrong, black-white, etc. without discriminating degrees between extremes. Beliefs are rigidly held. A person holding a different belief is regarded as an adversary. Data or facts not consistent with the belief are explained away or ignored.

When two persons (or groups) are both reactive, a chain reaction may occur between them. 'A' reacts angrily to something 'B' has done or failed to do. 'B' reacts to 'A's reaction, ignoring what 'B' had done or not done. 'A' then reacts to 'B's reaction, ignoring the fact that 'B' had reacted to 'A's reaction. And so on.

iii) *Uniordinal.* In this mode, a party may feel emotion without being driven by it, and can and does distinguish degrees between extremes. Another person is not regarded as totally good or totally bad, but as being good in some ways, bad in others. Both opposite extremes of a dispute are seen as having some merit, and a compromise between those extremes is sought. Beliefs are not so rigidly held, and are modified if an inconsistency is shown.

But awareness, while broader than in the Reactive Mode, is still limited. The individual views events from one perspective — his own, or that of the group to which he belongs. There is little or no empathy for other perspectives. There is a tendency to 'selfishly' pursue one's own goals with little or no regard for the effect of one's action on others.

iv) *Multiordinal.* In this mode, awareness expands to take into consideration the perspectives and values of others as well as one's own. There is increased empathy — the ability to put oneself in another person's shoes, and to understand how he feels and what he thinks, without necessarily approving or agreeing. The multiordinal individual is able to see how his action may affect the goals and interests of others. He is also able to view his own activity from more than one perspective — e.g., he can think about a problem while jogging.

v) *Synergic.* In this mode, an individual or party not only views situations and events multiordinally; he also focuses sharply on things that promote two or more goals, interests, values, etc. He can achieve a high degree of empathy with others, and readily modifies his actions so as not to impede the interests of others. His thinking and actions are usually multipurposed. He appreciates the value of synergy and seeks to promote it.

The Mode of Function Scale provides a basic perspective from which to promote synergy and reduce dysergy. Clearly, the Identic and Reactive modes tend to promote dysergy, so that it becomes desirable to de-activate identifications and reactions as much as possible. The Multiordinal and Synergic modes tend to promote synergy, so the more they can be facilitated, the higher the quality of life. The Uniordinal mode is in the middle, but in a highly interactive situation it tends to produce dysergy.

In applying the teleogenic system methodology, the change agent continuously and quietly appraises the system from the perspective of the Mode of Function Scale. He does not totally reject a goal or process just because it is reactively held; and he does not communicate his evaluation to the party involved (this would only provoke a reaction!) Rather, he seeks to enable the party to view the goal or process uniordinally. If and when this is done, he then seeks to evoke a multiordinal, and finally, a synergic mode orientation. He does not try to change the party's mind about the goal. The

TABLE 1
ACTIONS OF METASYSTEM CHANGE AGENT

Mode Detected	**Next Mode Desired**	**Type of Work Required**	**Reason**
Identic	Reactive	Contact	Enable awareness
Reactive	Uniordinal	Discrimination	Enable distinctions between extremes
Uniordinal	Multiordinal	Perspective	Enable expansion of awareness, empathy
Multiordinal	Synergy	Synergy	Enable synergy of goals, interests, values, functions, etc.

party *may* change his goal as he rises to the synergic mode; but that is for him to decide. A reactive belief may have some merit; usually it is the rigidity with which the belief is held or the reactive way actions are taken to achieve the goal, that causes dysergy.

A simple scheme summarizing the kind of work the change agent seeks to promote is given in Table 1. (For more details, see Coulter[22]).

5. *Nature of solution obtained.* Actuality theory, a clear understanding of the multiple purposes pervading a complex human system, and an ability to apply the concepts of synergy, dysergy, and the Mode of Function Scale, are part of the resources of the change agent. But he also carefully limits and focuses his actions and communications. He does not try to directly change another's mind, but rather *enables* change to occur, recognizing that teleogenic systems (individuals, groups, etc.) have uncontrollable components. He does not simply try to organize or reorganize the system per se; he creates new channels to promote synergy and avoids using channels that produce dysergy. He does not avoid conflicts, but seeks to develop a common actuality which enables the parties to view the conflict as *dysergy,* a problem both have a common interest in resolving. He seeks to provide safe outlets for emotional discharge of contending parties, until they reach a point where they realize the waste of human energy involved; when this is reached, he tries to introduce new perspectives in terms of which both parties gain. In a word, he seeks to change a 'Win-Lose' conflict into an 'All-Win' situation, and finally, he recognizes that all problems in complex human situations cannot be immediately solved with the resources available; some must simply be lived with.

6. *Criteria to evaluate 'truth'.* Does a process promote synergy? Does it reduce dysergy? These are the simple basic criteria to evaluate 'truth' in the teleogenic system methodology.

It is important for the change agent himself to be at least multiordinal in mode when applying these criteria. Hence self-honesty and a continuous quest for greater self-understanding must be part of the life style of the change agent.

7. *Proof of 'truth'.* Evaluative judgments should be rendered by an observer who is himself at least multiordinal at the time judgments are made — ideally by someone who is not the change agent, but who knows teleogenic system methodology, as well as other aspects of the metasystems paradigm. The basic criteria can be applied from various perspectives, such as:

a) Are the values, interests, and perspectives of all involved adequately considered and respected?
b) Is there an effective focus on promoting synergy among the communications, actions, and interactions of the parties?
c) Is there an active mutually acceptable procedure for confronting and eliminating dysergy?
d) Do initiators and leaders actively modify their actions to promote the goals and interests of others affected by those actions?

e) Are the 'rewards' of achievement synergic — i.e. do they not only benefit the achiever, but also enable him to contribute to the well-being of others?
f) Do parties have high empathy for each other?
g) Does 'think-feel synergy' — mutual facilitation of thinking and feeling — characterize the rational processes of the system?

Evaluative judgments made from these and other perspectives are not 'yes or no' judgments, but matters of degree. They are not made atomistically, but with consideration for the context involved — the other aspects, the various actualities, the situational stresses, etc. They are recognized as honest approximations which may be in error or otherwise inadequate, not as precise measurements with high accuracy. They are tentative and relative, not certain and absolute. And they are far more reliable than evaluations made with certainty by those who 'know' the truth.

REFERENCES

1. N. A. Coulter, Jr., 'Towards a Theory of Teleogenetic Control Systems', *General Systems,* **13,** 83-89 (1968).
2. N. A. Coulter, Jr., 'The Self-Determinism of Teleogenic Systems', *J. of Cybernetics* **5,** 9-20 (1976a).
3. N. A. Coulter, Jr., 'Contribution to a Triadic Theory of Development of Teleogenic Systems', *Proceedings of the Society of General Systems Research* **24,** 75-80 (1980).
4. N. A. Coulter, Jr. and A. G. Haddad, 'Algorithm for a Teleoid', *Proceedings of the Society for General Systems Research* **25,** 100-108 (1981).
5. M. LeGare, 'Overlapping Functional Systems: A Theory for Vertebrate Central Nervous System Function in Terms of Informal Systems Analysis', *Behavioural Science* **25,** 89-106 (1980).
6. A. Locker and N. A. Coulter, Jr., 'Recent Progress Towards a Theory of Teleogenic System', *Kybernetes* **5,** 67-72 (1976).
7. A. Locker and N. A. Coulter, Jr., 'A New Look at the Description and Prescription of Systems', *Behavioural Science* **20,** 179-190 (1977).
8. H. A. Simon, *Models of Thought* (Yale University Press, New Haven, Connecticut 1979).
9. J. P. Frisby, *Seeing: Illusion, Brain and Mind* (Oxford University Press, Oxford, England 1980).
10. P. M. DeRusso, R. J. Roy and C. M. Close, *State Variables For Engineers* (John Wiley and Sons, New York 1966).
11. J. Casti, 'Illusion or Reality? The Mathematics of Attainable Goals and Irreducible Uncertainties', *Int. J. of General Systems* **5,** 75-92 (1979).
12. S. Milgram, *Obedience to Authority* (Harper and Row, New York 1969).
13. L. Yablonsky, *Robopaths: People as Machines* (Penguin Books, Baltimore, Maryland 1972).
14. B. M. Gross, *Organizations and Their Managing* (The Free Press, New York 1964).

15. V. Papanek, *Design for the Real World Human Ecology and Social Change* (Ponthon Books, New York 1971).
16. G. Burrell and G. Morgan, *Sociological Paradigms and Organizational Analysis* (Heinemann, London 1979).
17. J. M. Burns, *Leadership* (Harper and Row, New York 1978).
18. J. Pfeffer, *Organizational Design* (AHM Publishing Corporation, Arlington Heights, Ill. 1978).
19. I. Illich, *Tools for Conviviality* (Harper and Row, New York 1973).
20. E. F. Schumacher, *Small is Beautiful* (Harper and Row, New York 1973).
21. J. P. van Gigch, 'A Methodological Comparison of the Science, Systems, and Metasystems Paradigms', *Int. J. Man-Machine Studies* **11**, 651-663 (1979). Chapter 2, this text.
22. *Synergetics: An adventure in Human Development* (Prentice-Hall, Englewood Cliffs, New Jersey 1976).

SECTION IV

METAMODELS AND APPLICATIONS

13. METAGAMES AND METASYSTEMS*

Rammohan K. Ragade

INTRODUCTION

The significance and relationship of the metagame approach to the metasystem paradigm are outlined. The foundations of the theory of metagames and recent development in the field are surveyed. Metalogical extensions and metarational outcomes are discussed, as they relate to the metasystem framework.

SUMMARY

I am fascinated to see the term metagames appear even in popular literature: Joseph Deken in his paperback book *The Electronic Cottage*[1] has a whole chapter titled: GAMES, HYPERGAMES AND METAGAMES. Now it appears possible for the arcades of video-games to be invaded by popular new metagames and hypergames! Finally an obscure relatively unknown branch of the theory of games has found a large scale practical application. It is true that within the artificial world generated by these micro-processor applications we could have any possibility!

Let us see what deep significance this concept has and what it indeed stands for. In our search for an understanding of the relationship of the metagame concept, to the metasystem paradigm,[2,3] it will be well to start with the basic concepts of game theory in as simple a manner as the theory will allow us to explain it.

The theory of games was originated by Von Neumann and Mórgenstern in their need to establish a common unifying approach to understand social and economic systems which are characterized by the preponderance of interacting decisions and personal preference patterns reflecting individual or group values or beliefs. Such a situation is not recognizable in the domain of other natural systems except at aggregate survival levels of species.

Decision makers are called players in the terminology of game theory, while the particular choices they can each make are referred to as strategies, tactics or policies. A preference pattern for the resulting outcomes range from qualitative terms to very definite measurable quantitative terms. Within this basic framework developed some rather constraining concepts of rationality, which were the basis of much philosphical speculation and mathematical analysis ranging from Kenneth Arrow's[4] famous paradox of rationality to the paradox of rationality in Hardin's example of the Tragedy of the Commons.[5] This approach constitutes the normative approach of decision theory, game theory or social choice theory.

While much empirical research evidence was also available to show many

* Adapted from R. K. Ragade *Complexities and Conflicts in Multipurpose Systems Design* (Department of Systems Design, University of Waterloo, Waterloo, Ontario, Canada, 1976). Permission granted by the University of Waterloo.

unexplainable results by the normative theory, the preponderance of non-quantative and non-transitive preferences in the field of international relations were baffling the game theorist. It is in this milieu that Nigel Howard devised the idea of a metagame which is a mathematical structure in which we can talk about a formal game as constituted by the theory of games and use the very same principles to investigate other properties through a metalogic procedure.

The metagame approach creates a metalanguage out of the elements of the given game. The theory of metagames, the methodology of analysis of options based on metagames and the concept of empirically and experimentally validated metagame theories has been principally the work of Nigel Howard.[6–11]

He developed this approach to game theory to analyze problems of escalation and de-escalation of international conflicts. The non-quantitative non-numeric approach chosen was motivated by the difficulty of assigning any kind of tangible and quantitative values to international relations. Preferences are usually stated by politicians in qualitative terms. The strategic considerations are finite in number although beset with a multiplicity of logical conditions. In the middle of the '60s, game theory was very heavily influenced by cardinal utility theory. Any type of meaningful analysis of real world problems, particularly in the area of international politics, by classical game theory (with immediate implications for decision-makers and chief executives of a nation) was a hopeless task. It was about this time that Peleg[13] pointed out the independence of game theory and utility theory.

Nigel Howard's approach seemed promising and, as he has reported in his book[9] and elsewhere,[10,11] it has been applied in the study of the Vietnam Conflict, the Arab-Israeli War and the SALT talks.[14] Other applications have been made within governments and businesses. In terms of theoretical developments, in addition to Howard, Hildebrand,[15] Kalevar,[16] Saaty[17], Ragade[18–23] and Bennett[24] have contributed many new aspects. Radford has shown applications to Management Information Systems.[25]

REVIEW OF THE PROPERTIES OF A BASIC GAME

A game is simply a listing of a finite set of strategies S_i for each player, followed by a listing of each player's preferences M_i between the outcomes of Ω obtained from the interaction of players strategies. In formal mathematical notation a game G is written:

$$G = (S_1, \ldots, S_n; M_1 \ldots M_n)$$
$$\Omega = S_1 x \ldots x S_n$$
$$M_i = \Omega \times \Omega \qquad i = 1 \ldots, n$$

Consider an outcome named $\bar{\omega}$ with the constituent strategies of player i named s^{-1} and the combined constituent joint strategies of all other players named $\bar{s}^{N-i}$. Now consider another outcome ω which differs from outcome $\bar{\omega}$ to the extent that player i has chosen a different strategy s^i but that all other players have retained their joint strategy. This first outcome is said to be a RATIONAL OUTCOME for player i and the player a RATIONAL player when any of the second type of outcomes are not preferred by the player to the outcome $\bar{\omega}$. This is written alternatively as:

$$(\omega,\bar{\omega})\varepsilon M_i \text{ or } \omega\varepsilon m_i\bar{\omega} \text{ or } M_i\omega \subseteqq M_i\bar{\omega} \text{ or } f_i(\omega) \leqslant f_i(\bar{\omega})$$

The last form is appropriate in the case of games with ordinal or cardinal preference patterns.

The set of all such Rational Outcomes for player i is denoted by $R_i(G)$. The set of EQUILIBRIA in the game are them obtained from an intersection of the sets $R_i(G)$:

$$E(G) = \bigcap_{i\varepsilon N} R_i(G).$$

If G is a two person zero sum game, $E(G)$ will be the single saddle point if a solution exists in pure strategies, or else it is empty. For non-zero sum games and n-person games $E(G)$ defines the Nash equilibria. Since $E(G)$ may be empty or since Nash equilibria are not Pareto-Optimal we consider mixed strategy solutions. The famous minimax theorem is in the theory of two person zero sum games. Elsewhere, mixed strategies are not so spectacular a concept. Furthermore, mixed strategies suggest that the only thinking process involved is that of a gambler, who decides on the roll of a die. The latter may be appropriate for certain games such as parlor games or learning games, but not for most-of-a-kind strategic policy decisions, where the environment is never the same. Furthermore, players recognize that other thinking processes are present in the situation.

RELATED APPROACHES

Von Neumann and Morgenstern[34] themselves recognised this shortcoming (see p. 100–101 of their book) in terms of the majorant and minorant of a game. These are games related to the basic game G, in that they are derived from it by special construction. In the *minorant of a game* G, written min(G) player 1 is required to make his choice. Player 2 then acts in full knowledge of player 1's strategy. Player 1 is here at a disadvantage. In the *majorant of a game*, written maj(G), player 2 moves first after which player 1 moves, fully aware of 2's choice. Thus it is player 2 who is at a disadvantage. Both maj(G) and min(G) offer the respective players a straightforward choice. RATIONALITY is clear in such games which we may call *contingency games*. However, our interest is in game G where there are no contingencies. How are we to construct each player's possible thinking process? Clearly maj(G) and min(G) indicate a direction. If we take rationality in a narrow sense to mean the best play or strategy in the game by each player, we shall have a problem on our hands. We have created it to supply one or the other player with information regarding actions his opponent might take! This gets very complicated and ill-defined in the N-person case ($n > 2$), so that the narrow sense rationality is, in turn, ill-defined.

On the other hand, if by rationality we imply the faculty of reasoning to be employed by each player we have a richer structure. In this sense, we take the game to be a problem to be solved by the players.[35] It may be an ill-defined problem as most real world conflicts are likely to be[36] with a resulting hierarchical structure. Is it possible to follow the logical reasoning that players may have in their minds? Perhaps. Our understanding of human thought processes has been greatly enhanced by the developments in cognitive psychology[37–42] and Artificial Intelligence[43,44] and now with the formation of the new discipline of cognitive science. The *Weltanschauung* of each player would perhaps determine the type of thought processes likely to be followed by them. In large games with many players, each player may attempt to reason and determine his strategy based on patterns that he recognizes. This information processing aspect, of the perhaps not so well defined game, is important for us. We would like to construct games of the majorant-minorant type through a formal logical process that each player may go through in his own mind.

H. A. Simon[45–47] writes about the meaning of actions in a problem-solving situation. The reasoning process is an attempt to generate all possible logical sequences of the problem-solving situation, with which a sequence of *actions* is associated. Actions are goal directed i.e. a subset of outcomes is identified as being desirable. When applied to a game situation we see it as the thinking process by which players sort out their individual *action plans* to reach desirable outcomes. Such reasoning is more likely to be of the *satisficing* kind than the optimizing variety of the narrow sense rationality.

The theory of metagames is of games derived from a given game G. In this it is a *meta theory*. That is it takes the game G as its object of study and seeks to interpret results in G in a more generalized framework. It is a metalogical approach.[48]. The approach in metagames is strongly reminiscent of the first order and higher order predicate calculus of logic.[49] Such a metalogical approach has been considered also in other communications.[50,51] It thus dovetails into the metasystem paradigm which is the essence of this book.

METAGAME CONCEPT

A *metagame* models a player's particular cognitive process about the other players' cognitive processes, such that the outcomes in the game are seen as reasonable. It is to be considered a metalogical extension, through which the bounds of rationality are seen reduced to the basic game. By considering the reasonableness in all possible cognitive interaction, a player is able to identify reasonable outcomes in the game.

In planning strategies based on each likely contingency, a player is said to be creating a policy to be followed in a game situation. Take the Prisoner's Dilemma Game (PDG) as an example. Here C is the 'cooperative' strategy and D is the 'defect' strategy of each player. The reason we consider PDG is that it has some markedly unique properties not possessed by any other 2×2 game. In Rapoport and Guyer's taxonomy[52] it is in a class by itself. Of course, the metagame process is not unique to PDG. In Figure 1 if each player only considers contingencies as they occur, he may resort to narrow sense rationality. This leads to the equilibrium outcome d as shown in Figure 2. As we have said in the previous section these

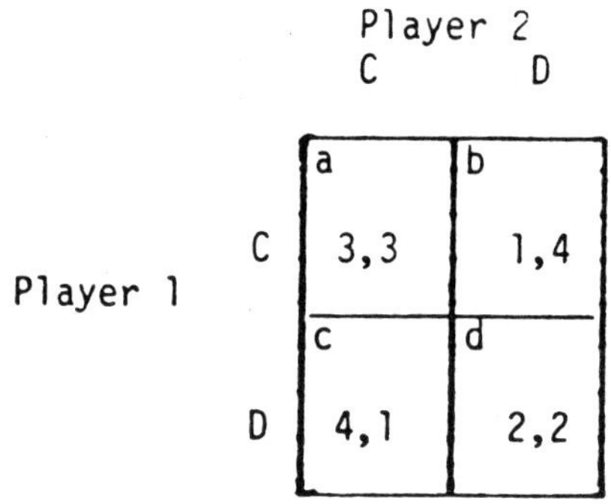

Fig. 1. The Prisoner's dilemma game.

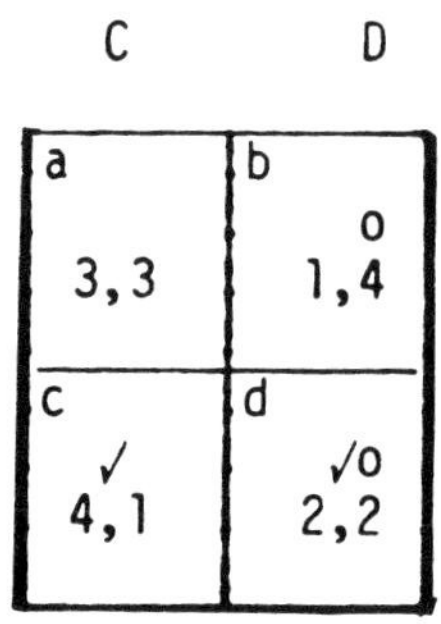

Fig. 2. Rational outcomes in the basic game G.

outcomes in Figure 2 are referred to as rational outcomes. The circle 'o' refers to outcomes rational for player 2 while the check mark '√' refers to outcomes rational for player 1. For the PDG, we shall frequently resort to referring outcomes in alternative forms for specific contexts. These equivalent forms are shown in Figure 3.

The equilibrium outcome (2,2) is rational for both players. However (3,3) dominates it. How are they to explain it? How will they reason it themselves? let us look at it from player 1's viewpoint. Suppose he chooses one out of the following contingency plans which we call *policies*:

(1) Always choose C, which we write: C/C

(2) Choose C when player 2 chooses C, and choose D when 2 chooses D: C/D

(3) Choose D when player 2 chooses C, and choose C when player 2 chooses D: D/C

(4) Always choose D: D/D

When these policies are matched against the strategies of player 2, we get a

a	(3,3)	(C,C)
b	(1,4)	(C,D)
c	(4,1)	(D,C)
d	(2,2)	(D,D)

Fig. 3. Outcomes in alternative forms for the PDG.

metagame. (Recall the majorant of *vN-M*). We call this game 1*G* and represent its outcomes as in Figure 4.

However 1*G* is not a complete game until we assign preferences to each outcome. For this, first we shall relate outcomes in 1*G* to outcomes in *G*. Then, we shall impart the same preferences to 1*G* as we have in *G*. In Figure 5, we show the relationship of outcomes in 1*G* to outcomes in *G*. Let us denote by Ω (1*G*) the former and Ω (*G*) or simply Ω the latter set.

Consider two outcomes z_1, $z_2 \varepsilon \Omega(1G)$. We shall say z_1 is not preferred to z_2 if their corresponding outcomes from Figure 5 are likewise not preferred. Hence for player 1, $\alpha \varepsilon M_i \lambda$ if and only if $a \varepsilon M_1 c$ which it is, where m_1 denotes the resulting preference relation. Likewise, for player 2, $\delta \varepsilon M'_2 \varepsilon$, if, and only if $d \varepsilon M_2 c$, which it is not. Hence $\delta \varepsilon M'_2 \varepsilon$. Let S'_i denote the policy set for player 1. Player 2 still considers S_2. Thus we have the object.

$$(S'_1, S_2, M'_1, M'_2)$$

	Player 2: C	Player 2: D
C/C	α	β
C/D	γ	δ
D/C	ε	θ
D/D	λ	μ

Fig. 4. Outcomes in 1G.

Ω (1G)	α	β	γ	δ	ε	θ	λ	μ
Ω (G)	a	b	a	d	c	b	c	d

Fig. 5. The β function corespondence.

which is a game. *Thus the metagame 1G is also a game based on G. It is a metalogical extension.* It has been developed primarily to improve the analysis of reasonable outcomes. By the metalogical process, we progressively reduce the *bounds on rationality* of choices faced in the basic game G. A study of all possible metagames and their commonalities will lead us to reasonable outcomes for strategic policy considerations. The metagame $1G$ can now be shown in full in Figure 6.

Thus the game kG is formed from the game G by replacing player k's strategy set S_k with the set of all possible functions f that can be defined from the joint *contingency* strategy set S_{N-k} (i.e. the cartesian product of all other player's strategy sets) to S_k. That is each such function f is defined as $f: S_{N-k} \rightarrow S_k$.

We call each f a *policy* or a *metastrategy*. The pair of (f, s^{N-k}) is an outcome in kG. To this pair an outcome in the basic game may be related which is $[f, (s^{N-k}), s^{N-k}]$. We call this relation the β function.

$$\beta\,(f, s^{N-k}) = [f\,(s^{N-k}), s^{N-k}]$$

We can now analyze game $1G$. In figure 7, an 'o' means rational for Player 1: a '\' means rational for Player 2, From the figure we have

$$R_1\,(1G) = \{(D/C, C), (D/D, C), (D/D, D), (C/D, D)\}$$
$$R_2\,(1G) = \{(C/C, D), (C/D, C), (D/C, D), (D/D, D)\}$$

C/C	α 3,3	β 1,4
C/D	γ 3,3	δ , 2,2
D/C	ε 4,1	θ 1,4
D/D	λ 4,1	μ 2,2

Fig. 6. The Metagame 1G for the PDG.

Player 2

	C	D
C/C	3,3	1,4 ✓
C/D	3,3 ✓	0 2,2
D/C	0 4,1	1,4 ✓
D/D	0 4,1	0 2,2 ✓

1G

Fig. 7. A Metagame 2G for PDG.

and hence

$$\hat{R}_1\ (1G) = \{(4,\ 1),\ (2,\ 2)\}$$
$$\hat{R}_2\ (1G) = \{(1,\ 4),\ (3,\ 3),\ (2,\ 2)\}$$
$$\hat{E}\ (1G) = R_1 \cap R_2$$
$$= \{(2,\ 2)\}$$

where the notation '^' is used to identify outcomes of the basic games that correspond to outcomes of a metagame. Alternatively, the notion of a 'β' function is used for this identification. Thus

$$\hat{R}_1\ (1G) = \beta R_1(1G),\ R_2(1G) = \beta R_2(1G)$$
$$\hat{E}\ (1G) = \beta E(1G)$$

When we refer to any outcomes in $\Omega(1G)$ as being rational then we only refer to $\Omega(1G)$. If however we wish to identify these rational outcomes of $1G$ in G, then we

Player 2

Player 1		C/C	C/D	D/C	D/D
	C	3,3	0 3,3	1,4 ✓	1,4 ✓
	D	0 4,1	2,2	0 4,1 ✓	0 2,2 ✓

Fig. 8. Rational outcome in 1G.

Player 1 \ Player 2	C/C	C/D	D/C	D/D
C/C/C/C	3,3	1,4	3,3	1,4
D/D/D/D	4,1	2,2	2,2	4,1
D/D/D/C	4,1	2,2	2,2	1,4
D/D/C/D	4,1	2,2	3,3	4,1
D/D/C/C	4,1	2,2	3,3	1,4
D/C/D/D	4,1	1,4	2,2	4,1
D/C/D/C	4,1	1,4	2,2	1,4
D/C/C/D	4,1	1,4	3,3	4,1
D/C/C/C	4,1	1,4	3,3	1,4
C/D/D/D	3,3	2,2	2,2	4,1
C/D/D/C	3,3	2,2	2,2	1,4
C/D/C/D	3,3	2,2	3,3	4,1
C/D/C/C	3,3	2,2	3,3	1,4
C/C/D/D	3,3	1,4	2,2	4,1
C/C/D/C	3,3	1,4	2,2	1,4
C/C/C/D	3,3	1,4	3,3	4,1

Fig. 9. The Game 12G.

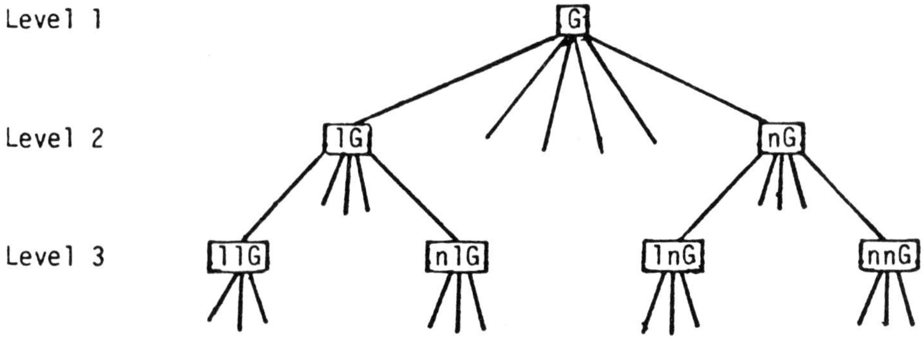

Fig. 10. A Metagame Tree.

use the β- function. The resulting outcomes are named *metarational outcomes*. In Figure 8 we show a PDG for $2G$.

$E(2G) = \{(D, D/D)\}$ and hence
$\hat{E}(2G) = \{(2,2)\}$

We can see the outcome (C,C) which dominates (D,D) is not the equilibrium in either $1G$ or $2G$. The metagames $1G$ and $2G$ are known as metagames in Level 1. We can formulate the game $12G$ from $2G$ (and $21G$ from $1G$), by recursion. The $12G$ is so constructed by supposing that Player 1 chooses his strategy after the knowledge of Player 2's choices in $2\ G$. We show this game in Figure 9. The metagames $12G$ or $21G$ are at Level 2. In general, we can construct a metagame tree for an n-player game as shown in Figure 10. The games such as $11G$, nnG etc. are for morphological purposes.

OUTCOMES IN THE METAGAME TREE

This metagame tree is referred to as the special metagame tree. Other metagames have been considered[11,12,20,21,24] and, indeed, may be considered by incorporating each persons beliefs, cognitive and judgmental processes. We shall restrict our attention to the special metagame first considered by Howard.

Characterization for an Arbitrary Metagame

An arbitrary metagame H can be written as:

$$k_1\ k_2\ .\ .\ .\ k_r\ G$$

where the players $k_1\ k_2\ .\ .\ .\ k_r$ are *named*, i.e. the reaction patterns are built on the decisions of players not in this list followed by $k_r\ k_{r-1}\ .\ .\ .\ k_1$ in that order.

Player k_1, reacts to players k_2's reaction to player k_3's reaction to player k_4's choice.

Definition: A *prime metagame* is one in which each player is named at most once.
Definition: The *prime representative* of a given metagame H is the prime metagame obtained by striking out all but the last occurrences, i.e., the one furthest to the right of each player's name from the title of H.
Example: Consider H = 513524312G. The prime representative of H is ~~5~~ ~~1~~ ~~3~~ ~~5~~ ~~2~~ 4 3 1 2 G = 4312G

Suppose that in the original game G there were six players. Then player 6 is said to be not named in the above metagame H. Players 1-5 are *named*.
Definition: Player k follows player j in $H(k \neq j)$ if some occurrence of k's name is further to the right of some occurrence of j's name in the title of H.
Definition: If k follows j in the prime representative of H, or if k but not j is named in H, then k is said to *follow j last*. For example, in the six player game G we have considered the metagame H = 513524312G. Then 5 follows 3 *and also* 3 follows 5. But, 3 follows 5 last.
Definition: A metagame H in which all the players are presented is called a *complete* metagame. This leads to the obvious definition of a complete prime metagame, and of a complete prime representative. Again for the example H = 513524312G is *not* a complete metagame.

We shall now state an important theorem in metagame theory, after some more notation is introduced.

Consider a metagame H, originally from an N player game, G. Let D be the set of players named in H. Let U be the set of players unnamed in H, $U = N–D$. Let F be the set of players who follow a player $i \varepsilon D$ last. Let $P = N–U–F(i)$ be the set of players who 'precede' i in the prime representative of H. Then the characterization theorem can be stated as follows:

Theorem 1 Characterization Theorem for metarational outcomes in arbitrary metagame H
A basic outcome $\bar{\omega}$ is metarational for player i from H if and only if:
There exists s^P, *for all* s^i, *such that, there exists* s^F where the outcome (s^u, s^P, s^i, s^F) is not preferred by i to $\bar{\omega}$.
or if the value function is used

$$\min_{s^P} \quad \max_{s^i} \quad \min_{s^F} \quad f_i(s^u, s^P, s^i, s^F) \leqslant f_i(\bar{\omega})$$

(The symbols 'V', '∃' are read, for all, there exists, respectively)

However, in most situations the metagame in which decisions are made is not available. We shall introduce special subsets of Ω which overcome this problem.

A study of arbitrary metagames in this context is relevant for designers of computer-based conversational systems. Metagames arise in the metacommunication processes of the users of such systems.[53]

An arbitrary metagame is therefore some hierarchic problem-solving structure. Since it is not always possible to know a specific metagame we shall derive conclusions from a family of its *descendant* of H: likewise H is the ancestor of kH.

Some corollaries may be derived from theorem 1 which are stated without proof (See Howard [10] for a proof).

Corollary 1 If H' is a descendant of H, then

$$\hat{R}_i(H) \subseteq \hat{R}_i(H')$$

(i.e.) the set of metarational outcomes are increased by considering a higher level metagame).

We have observed this in the example of Prisoners Dilemma Game discussed earlier where $R_2(G)r \subseteq \hat{R}_2(1G)$

Corollary 2 If $H' = kH$ then

$$\hat{R}_k(H') = \hat{R}_k(H)$$

(i.e. no new metarational outcomes are obtained by a player thinking and reacting to his own metagame).

Again this was observed in the same example where

$$\hat{R}_1(1G) = R_1(G).$$

The following two represent the same facets.

Corollary 3 If H^* is a prime representative of H

$$R_i(H) = R_i(H^*)$$

Thus there were no more metarational outcomes to be obtained at level 3 in the PDG.

Corollary 4 No more new metarational outcomes are obtained once all players are named.

Thus in a two player game we need consider metagames up to level 2 only.

General and Symmetric Metarational Outcomes

Let us denote by H an arbitrary metagame, perhaps of level r. Then by applying the β function r times to any subset P of $\Omega(H)$, the outcomes in H, we can identify the outcomes in G corresponding to outcomes in P. This relation we write

$$\hat{P} = \beta^r P$$

or when the level r is not known

$$\hat{P} = \beta^* P$$

so that β is applied sufficient number of times. Now consider the set

$$\Gamma_i(G) = \bigcap_H R_i(\hat{H})$$

over all possible metagames H. With the β function we identify

$$\hat{R}_i(H) = \beta^* R_i(H)$$

$\Gamma_i(G)$ is called the set of *general metarational outcomes* for player i. In the PDG, we have found that (3,3) is not a rational outcome in G, but it is rational in $12G$. Therefore we say (3,3) is metarational from $12G$.

Outcomes that are *metarational* from every complete metagame or equivalently from some descendants of every metagame are called *symmetric metarational outcomes*, which we denote by $\Sigma_i(G)$ for player i. The characterization of Γ_i and Σ_i are now given below. Let us recall the following symbols:

$R_i(G)$ is the set of rational outcomes for Player i in G

$\hat{R}_i(H)$ is the set of metarational outcomes for Player i from the metagame H

$$\Gamma_i(G) = \bigcap_H \hat{R}_i(H) \text{ is the set of metarational outcomes for Player } i$$

$$\text{[Note: } \bigcap_H \hat{R}_i(H) = R_i(G)\text{]}$$

We define $\Sigma_i(G)$ as the set of symmetric metarational outcomes for Player i (metarational from some descendant of every metagame)

$$\Sigma_i(G) = \bigcap_H \hat{\Gamma}_i(H)$$

Now

$$E(G) = \bigcap_i R_i(G) \text{ is the set of equilibria in } G.$$

$E(H)$ = the set of outcomes identified by the equilibria in H.

or $\bar{E}(H)$ = the set of basic outcomes that are obtained from the metagame H and correspond to the equilibria in H.

$$\text{[Note: } E(H) \subseteq \Omega(H)\text{]}$$

$\hat{E}(H)$ can also be expressed as

$$\hat{E}(H) = \beta^* E(H)$$

where β^* is the function to obtain the basic outcome, applied a sufficient number of times.

Likewise

$$\bigcap_i \Gamma_i(G) = \Gamma(G) \text{ is the set of general metarational equilibria}$$

$$\bigcap_i \Sigma_i(G) = \Sigma(G) \text{ is the set of symmetric metarational equilibria.}$$

All of these can be compactly written in Figure 11. To explain $\Gamma_i(H)$ note that $\Gamma_i(G)$ is defined for the node G in the metagame tree. Thus, from any descendent node $\Gamma_i(H)$ can be defined for the node H in the metagame tree, and $\hat{\Gamma}_i(H)$ are outcomes in G that correspond to meta outcomes $\Gamma_i(H)$.

Theorem 2. The set $\Sigma_i(G)$ can be characterized as follows:

$$\Sigma_i(G) = \{\bar{\omega} = (\bar{s}^i, \bar{s}^{N-i}) \mid \exists\, s^{N-i}\ \forall\, s^i: (s^i, s^{N-i})\ \varepsilon\ M_i\bar{\omega}\}$$

or

$$\Sigma_i(G) = \left\{\bar{\omega} = (\bar{s}^i, \bar{s}^{N-i}) \mid \min_{s^{N-1}} \max_{s^i} M_i\omega \leqslant M_i\bar{\omega}\right\}$$

in the case where $M_i\bar{\omega}$ can be treated as a value.

$R_i(G)$	$E(G) = R_N(G)$	$\bigcap_{i \varepsilon N} R_i(G)$
$\Gamma_i(G)$	$\Gamma(G) = \Gamma_N(G)$	$\bigcap_{i \varepsilon N} \Gamma_i(G)$
$\Sigma_i(G)$	$\Sigma(G) = \Sigma_N(G)$	$\bigcap_{i \varepsilon N} \Sigma_i(G)$

Fig. 11. The Sets R, Γ, Σ.

Theorem 3. The general metarational outcomes are characterised by

$$\Gamma_i(G) = \{\bar{\omega} = (\bar{s}^i, \bar{s}^{N-i}) \mid \forall s^i \; \exists \; s^{N-i} : (s^i, s^{N-I}) \; \varepsilon \; M_i \bar{\omega}\}$$

or

$$\Gamma_i(G) = \left\{ \bar{\omega} = \bar{s}^i, \bar{s}^{N-i}) \mid \max_{s^i} \; \min_{s^{N-i}} M_i \omega \leqslant M_i \bar{\omega} \right\}$$

where $\omega = (s^i, s^{N-i})$

The sets $R_i(G)$, $\Sigma_i(G)$ and $\Gamma_i(G)$ satisfy the following inequality.

$$R_i(G) \subseteq \Sigma_i(G) \subseteq \Gamma_i(G)$$

Explanation of the Characterization

For $\omega = (s^{-i}, s^{-n-i})$ to be symmetric metarational, we require the existence of some contingency s^{N-i} reached by the players N–i such that all outcomes in it, i.e. the set

$$\{\omega/\omega = (s^i, s^{-N-i}), s^i \varepsilon S_i\},$$

has outcomes not preferred in $\bar{\omega}$. In the case of a two player game this is easily seen by a matrix representation.

Any outcome in the column determined by the contingency $\hat{s}^{N-i}$ is not preferred to $\bar{\omega}$ by player i.

This contingency $\hat{s}^{N-i}$ is called a *sanction*. We use this concept in the metagame analysis algorithm for testing the stability of any outcome.

In the case of general metarational outcomes a specific contingency is not required. For each strategy s^i that is thought of improving ω for player i, i.e. may result in an outcome $(s_i, \bar{s}^{N-i})$ which is preferred to $\bar{\omega}$ the players N–i think of an appropriate contingency $\bar{s}^{N-i}$ *such that* (s^i, s^{N-i}) is not preferred to $\bar{\omega}$. Thus player i may have to start all over again from the new outcome.

The explanation of the various metarational outcomes are summarized:

Rational outcomes are outcomes that are metarational from every metagame. These are obtained without any planning or thinking metalogically. Equilibria generated from rational outcomes display the strongest type of stability. Players have to be induced to move from an equilibria.

Symmetric metarational outcomes are in some descendent of every metagame. In terms of the basic game at such an outcome a player p's unilateral improvement is held in check by sanctions one or other of the players could levy against player p. Equilibria based on these are stable when sanctions are credible.

General metarational outcomes are metarational in some metagame. However stability is not guaranteed and it has only weak checks against movements away from it. Equilibria based on these are likewise very weakly stable.

Coalitions in metagame theory are considered just as in the main body of game theory. In any game G, coalition preferences are based on individual preferences. Consider two outcomes $\omega_i, \omega_2 \varepsilon \Omega$ and a coalition C, $C \neq \phi$, $C \subseteq N$. We shall write $\omega_i \varepsilon M_C \omega_2$ if for some $i \varepsilon C$ $\omega_1 M_i \omega_2$. Thus only if $\omega_1 \notin M_i \omega_2$ for all $i \varepsilon C$ can we say $\omega_1 \notin M_C \omega_2$. That is if ω_1 is preferred to ω_2 by each coalition member we say the coalition C also prefers ω_1 to ω_2.

From this consideration we can proceed to characterize *coalition rationality* in the basic game G as well as *coalition metarationality*. In the basic game G the set of all outcomes rational for a coalition C, is

$$R_C = \bigcap_{i \varepsilon C} R_i$$

Likewise the set of symmetric metarational outcomes for a coalition C is

$$\Sigma_C = \bigcap_{i \varepsilon C} \Gamma_i$$

and the general metarational outcomes for C is

$$\Gamma_C = \bigcap_{i \varepsilon C} \Gamma_i$$

Up to now we considered special metagames, where policies were constructed based only on the strategies and reactions of the other players. In that sense they were *contingency metagames*. In addition we have metagames where each other's preference knowledge is taken into account in policy planning. That is, the evaluations and pairwise preferences of outcomes of all the other players is known to each player. The zero sum two person game is one such special case where automatically each player knows the other's preference. In constant-sum games likewise a subset of players knows what their complement gets. Such metagames, where the knowledge of others' preferences is available may be termed *metagames with mutual awareness*. They are also contingency metagames, but each player gives due consideration to the other's strategy.

An approach to metamodels of games for taxonomic purposes was considered by the author.[22] Other types of metagames are possible, where a metagame is constructed based on :

i) All possible subsets of the strategy sets of each player called *general metagames* by Howard.[9]

ii) Where preference relations are fuzzy and therefore call for the concept of fuzzy metarationality by Ragade.[21]

iii) Where the game itself is considered part of a larger game and therefore the rationality in the given game is restricted.[21,24]

iv) Where conversation among players should be considered.[53]

v) Where other modal logics and fuzzy logics are possible in the players minds leading to other types of predicate calculi.

It is this process of metamodelling that leads to all kinds of metagames. Thus, the metagame is not restrictive to the special metagame considered in this paper. In each case the resultant rational outcome in that metagame is related to the outcomes of the original game. The latter are then considered metarational.

SYSTEM FORMATION, METARATIONALITY AND METASYSTEMS

In systems theory the concept of systems formation is an important one. It seeks to address the question of how larger systems evolve from smaller systems? The theory of General Living Systems proposed by James G. Miller[54] postulates that each new higher level of a system, a functioning (living) system or organization must have its critical processes within healthy ranges of functioning. He further postulates a controversial number of 19 such critical processes for both the matter/energy processing and information processing. Others such as Gray,[55] Csany[56] and Ulanowicz[57] sought the critical conditions for the appearance of a higher level system formed from lower level units.

Now, it is clearly of interest, how higher level (living systems) are formed from lower level (living) systems, particularly when individual purposes and goals are taken into account. If the higher level system is metasystem for the lower level units, then it must provide the basis and mechanisms for metalogical processes for retaining these units within it. In terms of the theory of metagames, what is it that makes the metasystem appear metarational to each and every component unit? How stable is this within the logical processes of the components and the metasystem.[20] Should we consider metastability in yet a higher order system?

Interacting decisions of purposive systems when networking themselves into a metasystem provide another case for the consideration of metarational stability of their interacting choices.[23] More recent developments in the theory of metagames such as the concept of hypergame,[24] adaptation processes via metagames and fuzzy metarationality[20,21] may provide new insights into the formation of metasystems.

REFERENCES

1. J. Deken, *The Electronic Cottage* (Bantam Books, New York N.Y., 1981).
2. J. P. van Gigch, 'A methodological comparison of the science, systems and metasystems paradigms', *Int. J. Man-Machine Studies* **11**, 651–663 (1979). Chapter 3 in this text.
3. J. P. Gigch, 'Epistemological Questions Raised by the Metasystem Paradigm', *Int. J. Man-Machine Studies* **20**, 501–509 (1984). Chapter 4 this text.

4. K. J. Arrow, *Social Choice and Individual Values* (John Wiley and Sons, New York, N.Y., 1951).
5. G. Hardin, 'The Tragedy of the Commons', *Science*1243–1248 (Dec. 13, 1968).
6. N. Howard, 'The Theory of Metagames', *General Systems* **11**, 167–200 (1966).
7. N. Howard, *The Analysis of Options: A Computer-Aided Method for Analyzing Political Problems*', vol. II of a Report Prepared for the U.S. Arms Control and Disarmament Agency, (University of Pennsylvania, 1969).
8. N. Howard, 'The Arab-Israeli Conflict: A Metagame Analysis', Department of Systems Design (University of Waterloo, 1970).
9. N. Howard, *Paradoxes of Rationality: Theory of Metagames and Political Behaviour*, (M.I.T. Press, Cambridge, Mass. 1971).
10. N. Howard, 'Metagame Analysis of Business Problems' Part I: Description of the Technique **13**, No. 1, 48–67, (Feb. 1975), *INFOR*.
11. N. Howard, 'General Metagames' In: Anatol Rapoport (ed.) *Game theory as a Theory of Conflict Resolution* (Reidel Boston, Mass. 1975).
12. N. Howard, 'Examples of a Dynamic Theory of Games', Papers of *Peace Science Society International* **24**, (1975), 1–28.
13. B. Peleg, 'The Independence of Game Theory and Utility Theory', *Bull. Amer. Mathematical Society* **72**, 995–999 (1966).
14. L. Calhoun, 'Metagame Analysis: A Tool for Evaluating Negotiation Options with (SALT) Examples', Paper Presented at TIMS/ORSA Conference (Chicago 1975).
15. P. Hildebrand, 'Stable Outcomes in Two-Person Games', *Ph.D. Thesis* (University of Waterloo, Waterloo, Ontario 1973).
16. C. K. Kalevar, 'Metagame Taxonomies of 2 × 2 Games', *General Systems* XX, 195–210 (1975).
17. T. L. Saaty, *Mathematical Models of Arms Control and Disarmament* p. 82–111 (John Wiley, New York, 1968).
18. R. K. Ragade, 'On Some Games in a Finite State Sequential Machine Context', *Proceedings of the 16th Midwest Symposium on Circuit Theory* (Waterloo, Canada, April 1973).
19. R. K. Ragade, 'Transition Matrices of $FSSM_2$ Pregames' *Tech. Report*, March 27, 1973, Department of Systems Design (University of Waterloo, Waterloo, Ontario, 1973).
20. R. K. Ragade, 'Adaptations and Conflicts: A Game Theoretic View', *Proceedings of the International Conference on Information Sciences and Systems* (Patras, 1976).
21. R. K. Ragade, 'Fuzzy Games in the Analysis of Operations', *J. Cybernetics* **6**, 213–221 (1976).
22. R. K. Ragade, 'Lattice Theoretic Approaches for a Taxonomy of 2^n Games', *General Systems* XXI, 61–72 (1976).
23. R. K. Ragade, 'Mathematical Model of Approximate Communication in Information Systems', *Proceedings of the 22nd Annual Meeting of the Society for General Systems Research* (Denver, Colorado, 1977), p. 334–346.
24. P. G. Bennett, 'Towards the Theory of Hypergames', *OMEGA* **5**, 749–751 (1977).

25. K. J. Radford, *Managerial Decision Making* (Reston Publishing Co., Reston, Virginia, 1975).
26. R. K. Ragade and V. K. Handa, 'The Analysis of Options Approach in Collective Bargaining', *J. of Collective Negotiations in the Public Sector* **2**, No. 2, 173–187 (Spring 1973).
27. R. K. Ragade and V. K. Handa, 'Conflict Analysis in Construction Projects', *Proceedings of Project Management Institute, Eighth Annual Meeting* (Montreal, Canada, September 1976) p. 121–130.
28. D. Danielescu and K. W. Hipel, 'Metagame Analysis of the Grand River Basin Water Supply Conflict', Department of Systems Design (University of Waterloo, Waterloo, Ontario, April 20, 1972).
29. K. W. Hipel, 'Systems Science in Water Resources', M.A.Sc., Project Department of Systems Design (University of Waterloo, Waterloo, Ontario, August 1972).
30. K. W. Hipel, R. K. Ragade and T. E. Unny, 'Metagame Analysis of Water Resources Conflicts', *J. Hydraulics Div.*, American Society of Civil Engineers **10**, 1437–1455 (1974).
31. K. W. Hipel, Political Resolution of Water Resource Conflicts' *Water Resources Bulletin*, 813–827 (1976).
32. K. W. Hipel, 'Metagame Theory and its Applications to Water Resources Systems', *Water Resources Research* **12**, No. 3, 331–339 (1976).
33. R. K. Ragade, K. W. Hipel and T. E. Unny, 'Metarationality in Benefit–Cost Analysis', *Water Resources Research* **12**, 15, 1069–1077 (1976).
34. J. Von Neumann and O. Morgenstern, *Theory of Games and Economic Behaviour* (Princeton University Press, Princeton, N.J., 1946).
35. R. Banerji, *A Theory of Problem Solving* (American Elsevier Co., New York, N.Y., 1969).
36. R. Banerji, 'The Structure of Ill-Structured Problems', *Artificial Intelligence* **4**, 181–201 (1973).
37. R. C. Schank and K. M. Colby (eds.), *Computer Models of Thought and Language* (A. W. H. Freeman, San Francisco, CA, 1973).
38. U. Neisser, *Cognitive Psychology* (Appleton, New York, N.Y. 1967).
39. A. Newell, J. C. Shaw and H. A. Simon, 'Chess Playing Diagrams and the Problems of Complexity', *I.B.M.J. Res. Dev.* **2**, 320–335 (1958).
40. A. Newell and H. A. Simon, *Human Problem Solving* (Prentice Hall, Englewood Cliffs, N.J., 1972).
41. R. S. Wyer, Jr., *Cognitive Organization and Change: An Information Processing Approach* (Lawrence Erlbaum, Potomac, Maryland, 1974).
42. D. E. Rummelhart, P. H. Lindsay and D. A. Norman, 'A Process Model of Long Term Memory' In: E. Tulving and W. Donaldson (eds.), *Organization of Memory* (Academic Press, New York, H.Y., 1972).
43. B. Chandrasekharan, 'Artificial Intelligence: A Survey of a Decade of Progress', *Advances in Computers* 1974 (Academic Press, New York, N.Y., 1974).
44. E. B. Hunt, *Artificial Intelligence* (Academic Press, New York, N.Y. 1975).
45. H. A. Simon, 'On Reasoning about Actions' In: H. A. Simon and L. Siklossy, (ed.) *Representations and Meaning: Experiments with Information Processing Systems* (Prentice Hall, Englewood Cliffs, N.J. 1972).

46. H. A. Simon, 'How Big is a Chunk?', *Science B*, **183**, 482–488 (18th Feb. 1974).
47. H. A. Simon and L. Siklossy (eds.), *Representations and Meaning: Experiments with Information Processing Systems* (Prentice Hall, Englewood Cliffs, H.J. 1972).
48. G. Hunter, *Metalogic: An Introduction to the Metatheory of Standard First-Order Logic* (London, Macmillan, London 1971).
49. B. Elspas, K. N. Levitt, R. J. Waldinger and A. Waksman, 'An Assessment of Techniques for Proving Program Correctness", *Computing Surveys*, **4**, No. 2, 97–147 (June 1972).
50. D. Harrah, *Communication: A Logical Model* (M.I.T. Press, Cambridge, Mass. 1963).
51. P. Watzlawik, J. H. Beavin and D. D. Jackson, *Pragmatics of Human Communication: A study of Interactional Patterns, Pathologies and Paradoxes* (Norton, New York, N.Y. 1967).
52. A. Rapoport and M. Guyer, 'A Taxonomy of 2 × 2 Games', *General Systems* **11**, 203–214 (1936).
53. G. Pask, *Conversation Theory: Applications in Education and Epistemology* (Elsevier, Amsterdam and New York, 1976).
54. J. G. Miller, *Living Systems* (McGraw Hill Co., New York, n.y. 1978).
55. W. Gray, 'System Forming Systems', I. W. Reckmeyer (ed.) *Proceedings of the Twenty-Fifth Annual North-Amercican Meeting of the Society for General Systems Research* (Toronto, Canada, Jan. 6–9, 1981).
56. V. Csany, 'General Theory of Evolution'. In: R. K. Ragade (ed.) *General Systems* XXXVI, 73–94 (1981).
57. R. E. Ulanowicz, 'An Hypothesis on the development of Natural Communities' In: R. K. Ragade (ed.) *General Systems* XXVI, 111–133 (1981).

14. EPISTEMOLOGICAL HIERARCHIES IN KNOWLEDGE AND EXPERT SYSTEMS

Mildred L. G. Shaw

SUMMARY

Increasingly computers are being used as tools for aiding knowledge acquisition and decision making, either alone or in some form of man-machine symbiosis. This has culminated in the development of *expert systems* in which the computer acts as a medium for access to encoded expert knowledge. It is possible to give a system-theoretic framework for much of the operation of such computer-based systems enabling suitable algorithms to be formally derived. Most practical systems commence with some inbuilt vocabulary for the problem area for which they are designed and this itself is not subject to modification either by the computer or by the user. The author demonstrates a computer-based method of obtaining such vocabulary through the elicitation of personal constructs. This method allows the user to modify these knowledge structures in an interactive mode as they are obtained.

INTRODUCTION

Some degree of pre-supposed terminology is common to all problem-solving environments even when automation is not involved. If we take a problem to an 'expert' then we first of all have to express it in such a way that he can understand it. We may well find, however, that this task is itself the most difficult of all.

This paper argues first of all for the importance of this terminological level in any form of knowledge acquisition and decision making system, and shows its significance in terms of Klir's[1] *epistemological hierarchy*. This is related to Kelly's[2] work on *personal construct* systems and to Shaw's[3] techniques for *computer-based elicitation* of constructs and their analysis using Zadeh's[4] *fuzzy sets*. The operational value of this approach is demonstrated through an example of the computer program *PEGASUS* applied to a problem of eliciting the basic vocabulary and knowledge structures used in a defined problem area.

LEVELS OF KNOWLEDGE AND FUZZY SETS

One of the most useful frameworks within which to discuss knowledge structures is Klir's[1] *epistemological hierarchy*. Figure 1 shows the cycle of events→*modelling*→*prediction/decision/action*→*events* passing through this hierarchy.

The lowest level is one of *source systems,* effectively one of data terminology definition whereby the way in which any interaction with or discussion about a system of knowledge is defined and agreed. The next level is one of *data systems*, effectively one of basic knowledge acquisition, whereby the behaviour of some system is described in terms of the agreed domain of discourse at level zero. Levels above these

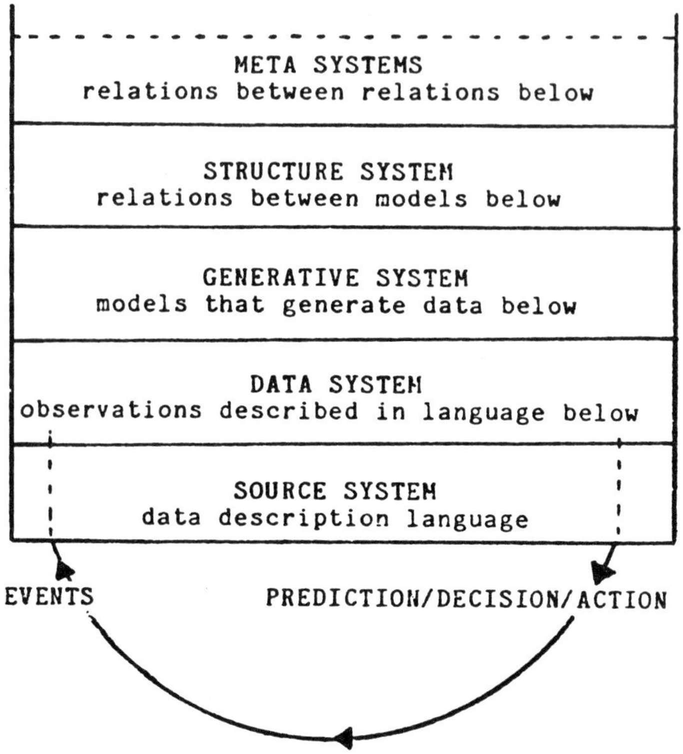

Fig. 1. The Epistemological Hierarchy (Klir[11]).

are concerned with models of data and their inter-relationships. Establishing any system of expert knowledge involves making a number of presuppositions at each level of this hierarchy.[5] From a formal point of view, the presuppositions taken at each level are quite distinct from one another and they are also quite arbitrary. From a practical point of view, however, the presuppositions at different levels interact with one another and are certainly not arbitrary since they determine the success of a specific expert system. If it is not explicitly realized that such presuppositions are being made and are necessary to any operational system then it is easy to become confused about the reasons for the success or failure of any particular system. It is also possible to generate fruitless debates about the merits of different approaches to decision making systems which are based on the failure to distinguish between presuppositions made at different levels in the hierarchy.

One presupposition that originates from the physical sciences and is often inappropriate to knowledge acquisition is that of *precision* in the source system and its application to create the data system. In the physical sciences the expected and preferred source system in which to represent data is one of exact measurement. We use a source system of physical quantities and their precise measurement and when we come to generalize this in knowledge acquisition we forget that the underlying constructs of physics have been desired and refined over a very long period and are themselves of a peculiar and, perhaps, unique nature. The existence of continuous and limitless scales for physical variables of length, time, mass, charge, and so on, is an important phenomenon that marks out the constructs involved as being different from those in many other sciences. However, the existence of refined measuring schemes for some constructs should not blind us to their close relationships to other constructs for which no such physical measurement exists. For example, compare the concepts of *tallness* and *beauty*. The concept, the perception, of *tallness* exists in a more primitive sense than does the measurement of *height*. We are able to generate and follow arguments involving *tallness* without having any concept of inches, centimetres, or any other metric scales. Whilst a *scientific* analysis might conclude that there is a wide and ill-defined range of physical phenomena that combine in an extremely complex fashion to produce the subjective impression of beauty, in everyday reasoning it is as primitive a term as tallness. We certainly do not distinguish between them in statements such as: 'this ladder is too tall to put in the kitchen'; and 'this vase is too beautiful to put in the kitchen'. We are capable of decision making such as: 'he likes girls that are tall and beautiful; Mary is not very tall but very beautiful; he will probably like Mary'.[6]

Fuzzy Set Theory [4,7,8] provides a model of constructs at the source systems level which is more primitive and generally more applicable than the quantitative scales of the physical sciences. Under certain conditions measuring schemes may be available which preserve all the order relations of the fuzzy sets and obey additional axioms. Deriving these additional, higher-order constraints, is itself an important task of empirical science and the general assumption at the source level can be only that there may be predicates relevant to the problems under consideration that define fuzzy sets in relation to the entities involved. The links between the basic constructs used at the source level and fuzzy set theory have already been established. Goguen[9] takes a formal axiomatic approach to the notion of a 'concept' in natural and artificial languages and shows within a very general category-theoretic framework that one obtains generalized fuzzy sets. Gaines[10] shows that the modelling techniques based on measures of complexity and approximation that are effective for stochastic data readily generalize to data that is both fuzzy and stochastic and leads to *possibilistic* models. Thus fuzzy set theory provides a syntactic basis for source and data systems that enables decision analysis to be generalized to the imprecise data of everyday life. There are, however, possible artefacts that arise from the *semantics* of the source and data levels and are often unrecognized as major determinants of system success or failure. The terminology used in describing data is tacitly accepted and its derivation is not taken to be a significant part of establishing the expert system. However, in the literature of both arts and sciences it will be found that descriptive terminology changes both organically and discontinuously as disciplines develop and undergo various forms of *revolution*.[11] The literature on knowledge acquisition and expert systems is most often concerned with the upper levels of Klir's hierarchy, such as the relationship between a data system and the generative system that models it, rather than the language in which the data itself is described. However, all of these upper levels, where refined modelling techniques may be applied, derive their significance, if any, only from the significance of the lower levels. No amount of techniques at the upper levels can compensate for lack of significant information expression at the lower levels. In the next section we discuss an approach to the operational analysis of these lower levels.

PERSONAL CONSTRUCTS AND FUZZY SETS

One of the most detailed studies of the operational role and significance of the source system level on personal decision making arises from the psycho-philosophical work of the clinical psychologist George Kelly. In his seminal work, *The Psychology of Personal Constructs*[2] he notes that:

> 'Man looks at his world through transparent patterns or templets which he creates and then attempts to fit over the realities of which the world is composed.' (pp.8-9).

Kelly emphasizes the epistemological role of these constructs in predicting and controlling the world and their ontological status as personal conjectures rather than reality-derived absolutes:

> 'Constructs are used for predictions of things to come, and the world keeps rolling on and revealing these predictions to be either correct or misleading. This fact provides the basis for the revision of constructs and, eventually, of whole construct systems.' (p.14).

This capability to revise construct systems in the light of experience and the pluralism of possible construct systems was the basis of Kelly's *constructive alternativism:* that we are not necessarily bound by our construct systems but can revise them to see the same problem from a new viewpoint. At a fundamental level we can see Kelly's *constructive alternativism* as being based on the primitive operation that Spencer Brown calls *making a distinction.* Brown commences his seminal work on *Laws of Form*[12] by noting:

> 'A universe comes into being when a space is severed or taken apart . . . The act is itself already remembered, even if unconsciously, as our first attempt to distinguish different things in a world where, in the first place, the boundaries can be drawn anywhere we please. At this stage the universe cannot be distinguished from how we act upon it, and the world may seem like shifting sand beneath our feet . . . Although all forms, and thus all universes, are possible, and any particular form is mutable, it becomes evident that the laws relating such forms are the same in any universe.' (p.v).

Thus, for both Kelly and Brown, all models at higher levels are fundamentally dependent on the distinctions made — the constructs used — at the source level. In designing expert systems we have a basic requirement to determine what these distinctions are, or should be, before proceeding with the analysis. Kelly terms this the *elicitation* of a construct system.

To show the significance of this in terms of the epistemological hierarchy, it can be redrawn with Klir's terminology of a *source* system replaced by Kelly's of a *construct* system.

Figure 2 shows this together with a more psychological interpretation of the upper levels of the hierarchy[13] which is useful in understanding an individual's acquisition

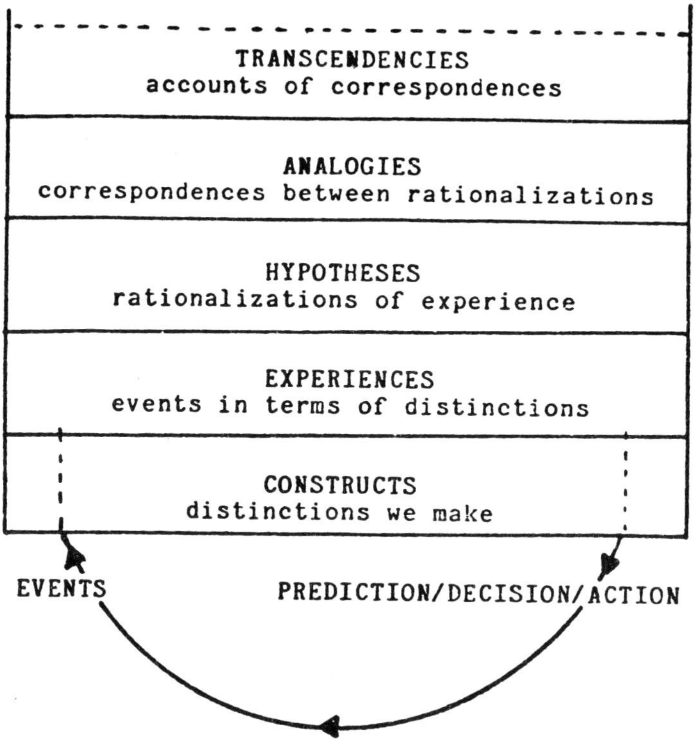

Fig. 2. The Epistemological Hierarchy (Gaines & Shaw[6]).

and application of expertise i.e. those of the personal scientist.[3] In this Figure the lowest level is one of the *constructs* or the *distinctions* made in interacting with the world. The next level is one of the *experiences* of events which happens to us, and we make happen, in terms of the distinctions already made. Levels above these are *hypotheses* which are rationalizations of experience, *analogies* between these rationalizations and *transcendencies*. Interaction with the world is, therefore, mediated through the construct system to produce experience which is modelled through the higher levels and leads to predictions, decisions and actions again mediated through the construct system.

Brown defined a distinction as a binary cut and Kelly also presented constructs as binary categories and based his own methodology for eliciting them on this. However, later workers found the need for shades of grey between the two poles of a construct and Kelly notes that this is consistent with his notion of a construct:

> "The construct, of itself, is the kind of contrast one perceives . . . while constructs do not represent or symbolize events, they do enable us to cope with events, which is a statement of a quite different order . . . They also enable us to put events into arrays or scales, if we wish."[14] (pp.13-14).

The interpolation within the basic bipolar distinction that has been made can be generated through logical operations that correspond to paradoxes of *self-reference* in classical logic. Varela[15] shows how a three-valued logical calculus arises from acceptance of a single self-referential form in Brown's calculus of distinctions. Gaines[6] shows how such 'primitive paradoxes' may be iterated to given an indefinite number of distinctions between the poles of those originally made, and hence how the fuzzy truth value of an arbitary proposition may be derived through a Dedekind section.

Kelly himself developed an elaborate but fairly informal epistemology based on his constructive alternativism and its consequences stated as a fundamental postulate and a set of 'corollaries'. He also developed techniques for eliciting the personal construct structure of particular individuals and it is this aspect of his work which has been primarily developed by later workers, mainly in the context of clinical psychology, that is, of people with abnormal construct systems. In recent years, applications of Kelly's ideas in management science has led to an interest in the personal construct systems of professional staff, and interactive programs have been developed that elicit an individual's construct system within a definite context.[3,16] Shaw and Gaines[17] have discussed such systems within the framework of general knowledge acquisition and the personal computer, and have given a fuzzy semantic model for the personal construct structures.

AUTOMATION OF CONSTRUCT ELICITATION — PEGASUS

To show how the methodology may be applied to the problem of the formation of a Citizen's Service Department (CSD), Shaw and Gaines[18] considered the attitudes of potential *Customers* to the *Actors* who currently carry out system activities related to the customer's address[19].

Figure 3 shows a repertory grid elicited from a person contemplating a change of address: the elements are different groups who should be notified and the constructs are significant distinctions between them. A '1' in the matrix indicates that left-hand description on the row applies to the element on that column; a '5' that the right-hand description applies; and intermediate values that the element lies between the two descriptions. This grid was derived by the computer program *PEGASUS* (Program Elicits Grids and Sorts Using Similarities)[3] which elicits a grid interactively from an individual, simultaneously acting as a psychological reflector for heightening his awareness and deeping his understanding of himself and his processes. This is done by continual commentary on related elements or constructs, together with the encouragement to differentiate between them.

Before choosing his elements, the user is asked to think about his purpose for eliciting the grid. Here it was stated to be 'who to inform when I change address'. This is of great importance for the interaction which is to follow, as it enables him to choose elements which are dependent on the purpose and then to distinguish between them through constructs which are relevant and representative of the topic area.

The *PEGASUS* program uses a standard triadic elicitation method to obtain the first few constructs as shown in the following dialogue:

		1	2	3	4	5	6	7	8	9	
authority	1	1	5	1	1	1	4	1	1	3	informal
government	2	5	3	5	1	1	5	5	1	5	commercial
takes money	3	5	3	3	1	1	2	4	3	1	gives money
unwelcome	4	4	5	3	1	1	5	4	4	5	welcome
main employment	5	1	3	3	1	5	4	3	3	3	consultancy
long-standing	6	3	3	1	1	5	2	3	1	1	fairly new
personal	7	3	1	5	5	5	5	5	3	2	impersonal
right to know	8	2	5	4	1	1	5	5	1	5	let know
see personally	9	1	1	1	5	2	5	5	1	1	remote
interferes	10	4	5	4	1	1	4	3	1	4	interacts
annoying	11	5	5	3	1	1	5	3	3	3	stimulating

9 automobile association
8 police
7 shareholdings
6 journals
5 vat authority
4 income tax authority
3 bank
2 friends
1 employer

Fig. 3. Grid elicited by PEGASUS from Person Contemplating a Change of Address.

```
TRIAD FOR ELICITATION OF CONSTRUCT 1
 1 employer
 2 friends
 3 bank

CAN YOU CHOOSE TWO OF THIS TRIAD OF ELEMENTS WHICH ARE
IN SOME WAY ALIKE AND DIFFERENT FROM THE OTHER ONE

WHAT IS THE NUMBER OF THE ONE WHICH IS DIFFERENT?2

NOW I WANT YOU TO THINK ABOUT WHAT YOU HAVE IN MIND
WHEN YOU SEPARATE THE PAIR FROM THE OTHER ONE.
HOW CAN YOU DESCRIBE THE TWO ENDS OR POLES OF THE
SCALE WHICH DISCRIMINATE employer AND bank ON THE
LEFT POLE FROM friends ON THE RIGHT POLE.

JUST TYPE ONE OR TWO WORDS FOR EACH POLE TO REMIND YOU WHAT
YOU ARE THINKING OR FEELING WHEN YOU USE THIS CONSTRUCT.

LEFT POLE RATED 1 -- ?authority
RIGHT POLE RATED 5 -- ?informal

NOW ASSUME THAT employer AND bank ARE ASSIGNED THE VALUE 1
AND friends IS ASSIGNED THE VALUE 5.

ACCORDING TO HOW YOU FEEL ABOUT THEM PLEASE ASSIGN TO EACH
OF THE OTHER ELEMENTS IN TURN A PROVISIONAL VALUE FROM 1 TO 5

 1 employer               1
 3 bank                   1
 2 friends                5
 4 income tax authority  ?1
 5 vat authority         ?1
 6 journals              ?4

POLE 1 --authority

 1 employer               1
 3 bank                   1
 4 income tax authority   1
 5 vat authority          1

 6 journals               4

 2 friends                5

POLE 5 --informal

DO YOU WANT TO CHANGE ANY OF THESE VALUES ? no
DO YOU WANT TO CHANGE THE POLE NAMES ?no
```

The *Construct Match* section of *PEGASUS* provides feedback when two constructs are highly related. The user is first asked to add an element which is either at pole "1" on the first construct and pole "5" on the second, or vice versa. If he can add a new element it must be rated on all the constructs so far elicited, but if he cannot split the two constructs this way, he is asked if he would like to delete a construct, combine two constructs into one, or just carry on. Feedback from a construct match is shown in the following dialogue:

```
THE TWO CONSTRUCTS YOU CALLED
 4 unwelcome--welcome
 10 interferes--interacts
ARE MATCHED AT THE 78 PERCENT LEVEL

THIS MEANS THAT MOST OF THE TIME YOU ARE SAYING unwelcome
YOU ARE ALSO SAYING interferes AND MOST OF THE TIME YOU ARE
SAYING welcome YOU ARE ALSO SAYING interacts

THINK OF ANOTHER ELEMENT WHICH IS EITHER
unwelcome AND interacts
OR interferes AND welcome

IF YOU REALLY CANNOT DO THIS THEN JUST PRESS RETURN AFTER
THE FIRST QUESTION MARK BUT PLEASE TRY.

THEN YOU MUST GIVE THIS ELEMENT A RATING VALUE ON EACH CONSTRUCT
IN TURN. AFTER EACH QUESTION MARK TYPE A VALUE FROM 1 TO 5
WHAT IS YOUR ELEMENT ?police

TYPE IN THE RATINGS FOR THIS ELEMENT ON EACH CONSTRUCT

LEFT POLE RATED 1, RIGHT POLE RATED 5

authority--informal ?1
government--commercial ?1
takes money--gives money ?3
unwelcome--welcome ?4
main employment--consultancy ?3
long-standing--fairly new ?1
personal--impersonal ?3
right to know--let know ?1
see personally--remote ?1
interferes--interacts ?1

DO YOU WANT TO CHANGE ANY OF THESE VALUES ?no
```

The *Element Match* section of *PEGASUS* provides feedback when two elements are highly related. The first choice offered is to add a new construct on which the two elements are placed at opposite poles, and then all the elements must be rated in the usual way. Alternatively, an element may be deleted, or no action may be taken. Feedback from an element match is shown in the following dialogue with *PEGASUS:*

```
THE TWO ELEMENTS 2 friends AND 6 journals ARE MATCHED AT THE 75
PERCENT LEVEL.

THIS MEANS THAT SO FAR YOU HAVE NOT DISTINGUISHED BETWEEN friends
AND journals

DO YOU WANT TO SPLIT THESE ?yes

THINK OF A CONSTRUCT WHICH SEPARATES THESE TWO ELEMENTS WITH
friends ON THE LEFT POLE AND journals ON THE RIGHT POLE

LEFT POLE RATED 1 -- ?personal
RIGHT POLE RATED 5 -- ?impersonal

ACCORDING TO HOW YOU FEEL ABOUT THEM PLEASE ASSIGN TO EACH
OF THE OTHER ELEMENTS IN TURN A PROVISIONAL VALUE FROM 1 TO 5
```

```
2 friends                  1
6 journals                 5
1 employer                 ?3
3 bank                     ?2
4 income tax authority     ?5
5 vat authority            ?1
```

When the grid reaches the maximum size allowed, or if the user chooses to finish before that, the results are analysed using the *FOCUS* procedure as shown in Figure 4. This is a two-way hierarchical cluster analytic technique which systematically reorders the row of constructs and columns of elements to produce a focused grid which shows the least variation between adjacent constructs and adjacent elements. The printout shows the trees of the elements and constructs as well as the focused grid with the element and construct labels. It can be seen from the element tree above the grid that the elements 'bank' and 'shareholdings' are seen as fairly similar despite the fact that this match appeared earlier in the interaction and the construct 'see personally — remote' was added to separate them. Also 'income tax authority' and 'vat authority' are seen to have similar characteristics. Looking at the construct tree on the right side of the grid three main clusters can be seen. Constructs 2,8,10,11 and 4 form a related cluster with 3, 5 and 6 less tightly linked, and 1,7 and 9 even more loosely grouped together. Since the *PEGASUS* feedback encourages the separation of constructs very close matches are rare.

As a *PEGASUS* elicitation proceeds, this *FOCUS* algorithm is used to offer to the user a possible explanation and interpretation of his meaning system in the terms of the similar patterns he uses in supposedly different circumstances. Cross-references are mapped across the grid and exhibited to the user in such a way as to offer him the facility to reconsider and change anything he feels to be inappropriate, which in turn enables him to be more aware of the links he is implicitly holding in his cognitive model.

GROUP CONSTRUCTS AND RELATIONSHIPS IN THE COMMUNITY — SOCIOGRIDS

As we begin to examine multiple construct systems, the duality becomes apparent between Kelly's extremely individualistic model of personal reality and our social stereotype of a common world shared through common ideas and language. Shaw and Gaines[20] have put forward a detailed model of how individual construct systems interact to form social construct systems, and how in doing this the individual systems are themselves changed in a way that models normative social influences. This goes a long way towards bridging the gap between the abstract epistemological hierarchy analysed by Klir and the pragmatic soft systems methodology of Checkland.[19] Again the methodology we propose has been at least partially operationalized through the program *SOCIOGRIDS*[3,16] which produces a set of *socionets* showing the capability of one person in a group to comprehend the constructions of the others when they have all construed the same elements. It also extracts from the group a set of *mode constructs* which are shared and hence form the basis of communication, consensual cognition and action amongst members of the group. *SOCIOGRIDS* has been applied in

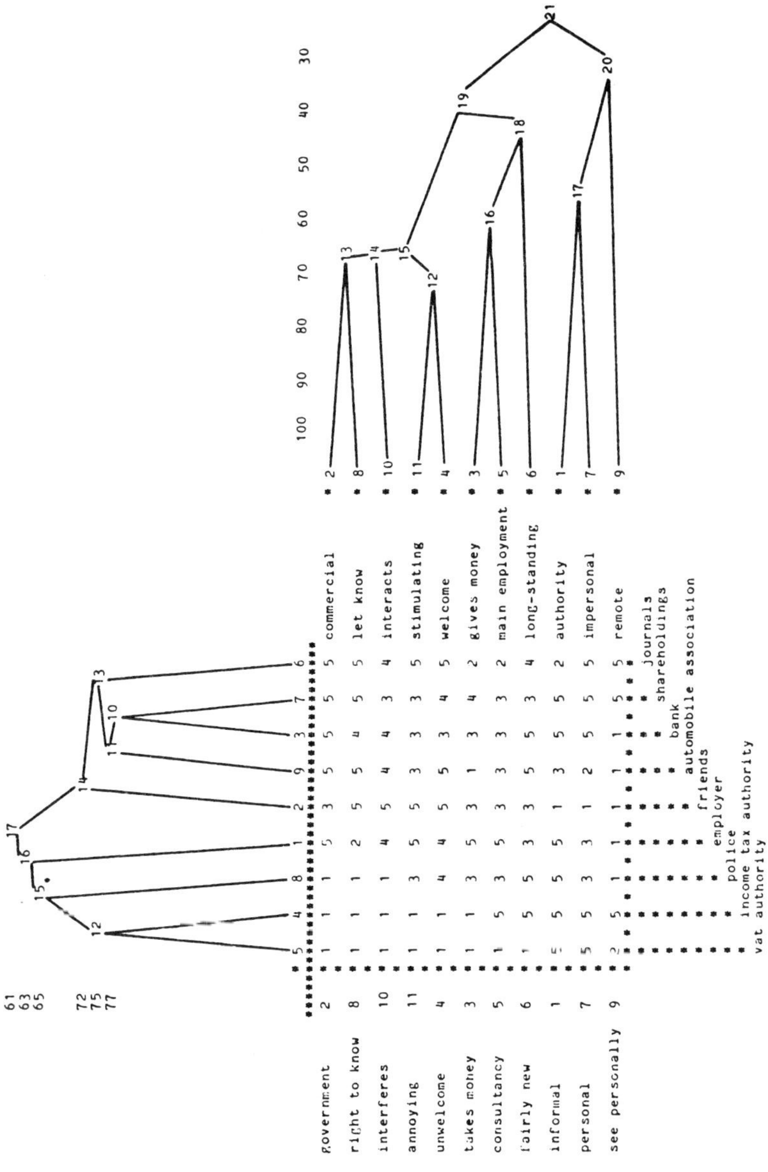

Fig. 4. Grid sorted by FOCUS to give a fuzzy hierarchical cluster analysis.

industrial studies of managers' evaluations of their subordinates and of the construct system of a quality control division in garment manufacture.[13]

What has been achieved through the interaction of an individual with the *PEGASUS* program described in this paper? From the starting position of stating the context as merely 'who to inform when I change address', we have come to a second stage where we have elicited: firstly, a set of organizations which are thought of as having to be informed; and secondly, a set of constructs that this person used in thinking about these organizations. We have done this in such a way that the person is stimulated to think about the variety of different organizations and the variety of ways he thinks about them. We have an *exploratory* program that itself has *no preconceptions* about the situation being explored — that is, one that begins to elicit the *whole* truth and *nothing but* the truth within that person's reality.

Even without any further analysis we have derived from the initial starting position a set of relevant aspects and ways of thinking about the problem. The clustering of the grid through *FOCUS* begins to give us *taxonomies* for these relevant features of the problem. By eliciting constructs from a wide range of individuals we can build up a fairly exhaustive repertoire of material on which to base our systems analysis. In the context of the CSD problem we might look at the similarities and differences between the construct systems of people in different sections of society using *SOCIOGRIDS.*

Figure 5 shows the role of the programs in relation to the problem representation for each of those playing roles in the root system and for the relationships between their construct systems. Note that Checkland's classification[19] brings into play a further hierarchical construct system based on the classification we use of people in society, and this classification can itself be explored using *PEGASUS* and *FOCUS.* Thus the methodology is truly *reflexive* and gives us the means to model ourselves and others exploring the means to model ourselves, themselves and others.

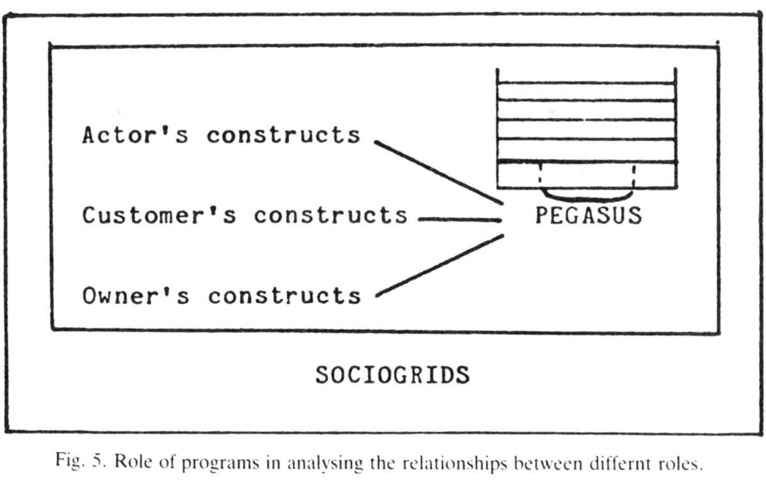

Fig. 5. Role of programs in analysing the relationships between differnt roles.

CONCLUSIONS

This paper argues for the key role of the lowest level of the epistemological hierarchy in knowledge acquisition. The source system of distinctions or constructs is a fundamental determinant of the operation, success or failure of all higher levels. It shows how the constructs being used at this lowest level may be elicited from a person through an interactive computer program, *PEGASUS.* It links the techniques described here with Checkland's soft systems methodology and the extraction of root definitions.[19] The paper also briefly indicates how the techniques can be applied to the multiple construct structures of groups and linked this with various psychosocial theories.[20] Programs related to *PEGASUS* have been developed for the analysis of group data such as *SOCIOGRIDS,*[3] which produces a *social network* based on mutual construing together with a *mode grid* of consequential construing. Thus theory and practice are bridged in providing techniques for knowledge representation and acquisition in expert systems.

ACKNOWLEDGEMENTS

I am indebted to my colleague and husband Brian Gaines for his help in the preparation of this paper.

REFERENCES

1. G. J. Klir, 'Identification of generative structures in empirical data', *Int. J. General Systems* **3,** 89-104 (1976).
2. G. A. Kelly, *The Psychology of Personal Constructs* (Norton, New York 1955).
3. M. L. G. Shaw, *On Becoming a Personal Scientist* (Academic Press, London 1980).
4. L. A. Zadeh, 'Fuzzy sets', *Information and Control* **8,** 338-353 (1965).
5. B. R. Gaines, 'System identification, approximation and complexity', *Int. J. General Systems* **3,** 143-174 (1977).
6. B. R. Gaines, 'Foundations of fuzzy reasoning', *Int. J. Man-Machine Studies* **8,** 623-668 (1976).
7. M. M. Gupta, G. N. Saridis and B. R. Gaines (eds), *Fuzzy Automata and Decision Processes* (North Holland, New York 1977).
8. E. H. Mamdani and B. R. Gaines (eds), *Fuzzy Reasoning and its Applications* (Academic Press, London 1981).
9. J. A. Goguen, 'Concept representation in natural and artificial languages: axioms, extensions and applications for fuzzy sets', In: E. H. Mamdani and B. R. Gaines (eds), *Fuzzy Reasoning and Its Applications* Academic Press, London (1981) pp 67-115.
10. B. R. Gaines, 'Sequential fuzzy system identification' *Fuzzy Sets and Systems* **2,** 15-24 (1979).

11. T. S. Kuhn, *The Structure of Scientific Revolutions* (University of Chicago Press, 2nd. Ed. 1970).
12. G. S. Brown, *Laws of Form* (George Allen and Unwin, London 1969).
13. B. R. Gaines and M. L. G. Shaw, 'A programme for the development of a systems methodology of knowledge and action', W. J. Reckmeyer (ed.), *General Systems Research and Design: Precursors and Futures* (Society for General Systems Research, Louisville, Kentucky 1981) pp. 255-264.
14. G. A. Kelly, 'A brief introduction to personal construct theory', In: D. Bannister (ed.), *Perspectives in Personal Construct Theory,* (Academic Press, London 1970) pp. 1-29.
15. F. J. Varela, 'A Calculus for Self-Reference', *Int. J. General Systems* **2,** 5-24 (1975).
16. M. L. G. Shaw, *Recent Advances in Personal Construct Technology* (Academic Press, London 1981).
17. M. L. G. Shaw and B. R. Gaines, 'Externalizing the personal world: computer aids to epistemology', In: *Improving the Human Condition: Quality and Stability in Social Systems* (Society for General Systems Research, Louisville, Kentucky 1979) pp. 136-145.
18. M. L. G. Shaw and B. R. Gaines, 'Eliciting the real problem', In: H. Wedde (ed.) *Model Realism* (Academic Press, London 1983).
19. P. Checkland, *Systems Thinking, Systems Practice* (Wiley, Chichester, U.K. 1981).
20. M. L. G. Shaw and B. R. Gaines, 'The Personal scientist in the community of science', In: W. J. Reckmeyer (ed.), *General Systems Research and Design: Precursors and Futures* (Society for General Systems Research, Louisville, Kentucky (1981) pp. 59-68.

15. REPLACING MODELS BY ACTORS; SOME IDEAS FOR METADESIGN

Gerard de Zeeuw

SUMMARY

In this paper a specific class of problems is considered: the problems that arise simultaneously with the implementation of a solution to another problem. This class of problem is labelled *PS-problems* (post solution problems). An interpretation is suggested: such problems appear when there is a variety of contexts to solutions, neither of which is dominant over time. Hence, such problems depend on a combination of two levels. One is the *local level* for the original problem, the other the *metalevel*, where the variety of contexts is defined. Next, the standard paradigm for research is interpreted as a one-actor model. A two-actor model for inquiry is then contrasted with this model. Specific characteristics of the two-actor model are derived, and it is explained how this model avoids the implementation of solutions leading to problems of the PS-type. Furthermore, an important characteristic of the two-actor model is that local worlds and meta-worlds (or levels and meta-levels) are not hierarchically ordered. That is, one may not learn to distinguish one or the other as being 'meta'. The two-actor model can therefore be interpreted as a form of meta-design: design based on the fact that actor-designers are producing something via each other's design. The paper describes how to acquire context-dependent knowledge, i.e. knowledge addressable to actors engaged in interaction with other actors, and thus, that build up each other's history.

THESIS

More so than ever, it seems that the word 'problem' (or its local equivalents), is one of the key words of our contemporary societies. There are personal and relational problems; economic problems; problems of pollution and resource depletion, etc. To understand this phenomenon some people point to the incredible complexity of most parts of our societies: schooling systems, transport systems, industrial complexes. Such complexity seems to have outrun our human abilities. Indeed, some tend to blame our lack of intelligence, and lack of skills for not being able to tackle all our problems more successfully.

In this paper I will assume neither of these positions. It is my thesis that the frequent occurrence of 'problems', signals a growing awareness that people can change their environments and societies—even though there may be obstacles to doing so. However, it is apparent that such obstacles no longer have to represent permanent barriers or constraints to such changes as was thought formerly. And thus, the obstacles have now become problems. The very frequency with which 'problems' arise is taken as a (positive) measure of a growing human ability to cope with change.

Starting from this thesis, methods to tackle some of the problems more effectively are derived and presented.

A special class of problems

The concept of *a problem* clearly is associated with that of *a solution* or *solutions*. However, solutions mean 'doing something'. Therefore, to call some situation problematic, implies an invitation to do something about it, as well as the assumption that a problem-solver will accept the challenge to tackle it. This explains the first part of the thesis. For indeed, in earlier times the assumption that a problem-solver existed seems to have been lacking, in the sense that people were assumed to be born into and thus to have a certain role and to stay within that role.[1] It follows, presumably, that there should have been, and were, fewer occasions to speak about *problems*. In this sense, it seems that nowadays people are assumed to be more active and involved than heretofore.

In addition, they have become also more ingenious. New solutions appear all the time. For example, people started first to organize labor as a discipline, solving the problem of how to handle complex machines. Second, they introduced the idea of holidays, as a way of escaping the discipline. Third, they found out that holidays were quite taxing—and the therapies were introduced to prepare for the holidays.[2] Similarly, rules of etiquette were introduced to help people make a choice between what they want or not to do, and to avoid coming into conflicts with what others want to do. However, these rules have also introduced the problems of loneliness and alienation—for the 'cure' of which, changes in social structures are sometimes advocated. Thus many of the problems people face, seem to be caused by the (earlier) introduction of solutions. In fact, we may even define such problems as a special class of problems: *post solution problems*, or *next-event problems*. Such problems can be clearly distinguished from other problems, like those of finding a solution to a differential equation where a well-defined solution is assumed to exist for each differentiation, or of designing tools given some design criteria. Here, the problems stem from the introduction of certain constraints, which allow for the possibility of a precise and, at least, 'satisfactory solution'. In the other cases, such 'satisfactory solutions' turn out to be less so, they are usually accompanied by post solution problems.

The defining and characteristic event here seems to take place when a solution to a problem is implemented. The question is: can one move to some other—mostly unrelated—problem, or is one forced to stay with certain accompanying problems—just to maintain that the earlier action can, at least for some acceptable span of time, remain a solution.

From the fact that it is so easy to indicate problems which are *post solution problems*, one may surmise that the latter class is quite extensive.[3] Therefore, it seems a reasonable to assume that a) over time, people have been able to introduce quite a few solutions, and that b) they show an overwhelming ingenuity to extend the class of *post solution problems*.

In summary, my thesis argues that:

There still remains a great deal that can be done to improve society systematically.

Post-solution problems can be differentiated from other types of *problems*, and,

The problem to solve *post solution problems* may introduce one more element to the class of *post-solution problems!*

INTERPRETATIONS OF THE DISTINCTION

There are several ways to interpret the existence of the *post solution class of problems* (PS-class). One, is to try to characterize the various possibilities of dealing with PS and non-PS problems. Another, is to try and find historical reasons for the production of such classes and of their relation. Both methods will be followed here. Other interpretations may be possible. However the ones mentioned seem to offer the best opportunity to further the development of approaches to the PS-class.

Let us first consider some examples that clearly do *not* belong to the PS-class. Suppose one defines some formal language. This again defines a class of problems, e.g. to prove that certain statements belong or do not belong to that language. Examples are algebras, or algebraic equations, computer languages, etc. Characteristically, such problems can be solved by means of the decision procedures defined by the language. As such, they usually are solved without creating other problems in their wake. One may say that this happens due to the fact that the language defines a unique *context*—or if one prefers, a *context* that dominates all other contexts, and makes the latter unnecessary to consider. For indeed, it is by dint of this uniqueness that we may say that such problems are non-PS. This property implies directly that solutions do not create problems; for, if they do, they cannot even be accepted as potential solutions.

Examples are not confined to formal languages. Another example is the work of Semmelweiss.[4] He found that in some wards of his hospital puerperal fever ran high. His problem became to reduce the incidence of this disease. After a long search, he reduced the number of possible contexts to one: doctors, that first worked on deceased people and then with pregnant women, were having a higher incidence of infection with puerperal fever than others. This provided a dominant context for solving his problem that became: prevent coupling of corpses and women, the cross-infection from corpses to women, either by sterilizing hands or by using different doctors for the two activities. No other context was needed to formulate a permanently non-problematic solution to the problem. Other contexts (location next to the mortuary, visibility of priests, etc.) did not provide such an opportunity.

Now consider some examples of PS-type problems. Suppose somebody asks for help, complaining about noisy neighbors. It may be possible here to find some presumably dominant context, like paper thin walls. A solution would be to raise the sound-damping quality of such walls. Will this be all? It may not be, as the noise producing neighbour may have been a noise lover, who also liked his or her neighbour's noise. He or she may now have a problem, given this new context. The wall-context thus was not clearly dominant, at least not in this case; the solution led to some PS-problems.

Similar consequences can be found in the escalation of the arms-race. Each solution to the problem of lagging behind, makes the other opponent unhappy, and so generates a problem that needs another solution. One may also think of traffic rules. These are solutions to traffic problems that, at one time, may have been considered adequate in a dominant context. However, increasing traffic load very often introduced new contexts, making the original one non-dominant and led to the need for traffic lights, highways, etc., introducing among others, problems of high expense, and of landscape change.

This last example introduces some elements with a historical interpretation. Some problems appear to be at first non-PS, due to the existence of a dominant context. Later on, other contexts may appear, calling for solutions in other dominant contexts, thus making the originally dominant context non-dominant. Thus, new problems appear: over time, the original solution creates PS-problems.

Two types of history are at play. One is that (dominant) new contexts just appear – e.g. because, as population increases, people can afford more traffic, new technology appears, etc. The other one is that the solutions themselves may be such as to create new and different contexts, where one or many of them dominate over what was originally thought to be the uniquely dominating context. An example is the (loneliness) problem facing Adam and Eve, solved by the advice: 'go forth and multiply'. The implementation of this solution was, in itself, influential in creating new contexts, with new problems[5], like those of overpopulation, and, in turn, of alienation.

The above interpretations were given to elucidate the relation between PS-problems and some of the non-PS problems. The question may be one of being able to find a domination context, or of creating non-dominance given an originally dominant context. Such a process may itself be started by finding a dominant context. In this case, we may speak of a self-destructive dominance, leading to PS-problems. In particular, this latter type of dominance will be considered here, including the most systematic reason for the existence of PS-type problems.

ONE-VERSUS MULTI-ACTOR INQUIRY

There are of course many ways of solving problems: by luck, by delegation, by some kind of conscious reflection and testing for adequacy, etc. We are interested here in systematic procedures, which lead to the formal acceptance of a proposed solution as a real solution. These procedures will be research procedures when one of the criteria used to accept a solution is that it be generalizable, i.e. enduring in a social, spatial and temporal sense.

A characteristic feature of such research is that it be *one-actor research* or inquiry as explained in what follows. A researcher—as an actor—is supposed to observe the world, as his or her environment and to 'abstract' as a solution to some problem some interpretable set of phenomena. 'Interpretable' here is taken to mean that it is possible to explain why a particular set is chosen, as well as the connection between the elements in terms of a dominant context. Of course, not all abstractions have the same or high quality. High quality is reached when the researcher can be (repeatedly) replaced by somebody else, who, in turn, will abstract in the same way as the first researcher. This replacement procedure actually is a testing procedure: will another researcher, given different available contexts, choose the same dominant context as previous researchers? The replacement procedure is quite standard. It will be clear from the above formulation why this procedure is called *one actor inquiry*. If it can be argued that different actors produce the same abstraction as if there were only one researcher, it is to be taken as a guarantee for quality.

It is not claimed here that all researchers apply such a one-actor model in their work. What is argued is that most descriptions of what researchers do, imply such a model, from which then, various types of research procedures are derived.[6] These range from isolation procedures to distinguish the set to be abstracted, to procedures to ascertain the degree of belief one may attach to certain statements concerning these sets. Following such procedures implies that certain aspects of the research output are stressed. The main one is emphasis on the existence of a dominant context, within which results—solutions to problems—remain solutions, and are not allowed to stray and produce new problems.

From the above formulation it can easily be deduced that there must be a class of situations where PS-type problems will occur. This happens in the case where, by accident, the dominance of a context disappears, or when the wrong dominant context is assumed. An example may be the situation where there are several actors, with one of them trying to act as researcher. To interact with the others, he or she may be interested in abstracting some regularity (in terms of a dominant context) in what the other actors do. However, the others may resist the necessary dominant context, and introduce other contexts, and produce PS-problems. For example, they may introduce some actions that will punish the researcher when he or she tries to find the regularity, or to use it. Thus, this will produce negative side-effects and therefore PS-type problems for the researcher. In situations like these, the one-actor model for inquiry, by the very process of being applied, contributes to the generation of such problems. As a result, the idea suggests itself to consider two-actor models for inquiry (or may be even multi-actor models).

Let me consider first the two-actor model. This may be formulated as follows. First the one actor or researcher, who is planning his or her action, must be introduced. Second, this actor will have to assume that, behind at least some of the phenomena to be considered, there is another actor/researcher, producing them. The central point here is that this second actor is assumed to be at least somewhat similar to the first. We may say that the world of phenomena is approached by the first researcher as if it is anthropomorphic, or better 'researcher-morphic'. The two researchers are *not* considered to be distinguishable as first or second actor.

The crux of this construction is that none of the actors can really 'abstract' as in the case of the one-actor model for inquiry. Indeed, each abstraction may be invalidated by the second actor, producing some change of context, possibly by his or her own intention to abstract.

It still may be possible to find some regularities for each of the actors, without having to introduce PS-type problems. This is the case when, at the outset, both actors *together produce* a dominant context. As long as this production is maintained, PS-type problems, for the solutions within this context, may be prevented from occurring. It must be noted that the production itself may still produce such PS/type problems (but they are prevented on a more local level).

In the next paragraph some consequences of the two-actor model, and some research procedures that are adequate to it will be indicated. First, one more general aspect of this question is discussed.

In what may called the *abstraction strategy* (for the one-actor model) the desired product is some '*model of*' the phenomena.[7] This is a picture or representation. Thus, when there is a second actor, whom we 'abstract' in terms of a set of behavioural phenomena, supplying the one-actor model, we replace this actor by a representation or '*model of*' some of his or her behaviour. Having done so, he or she no longer is 'necessary': his or her ability to introduce new contexts is assumed to have been subsumed under the dominance of the context in which the '*model of*' is constructed and justified. When next, we introduce the second (research) actor, in the sense of the two-actor model, we must again *replace the model by the actor*. Or more precisely: we assume in the two-actor model that the modelled actor has some similarity to the modelling actor, especially concerning the modelling activity, and the ability to generate new contexts. That is, the modelled actor should be able to start to produce the phenomena, which can be modelled as a '*model of*' the actor, via the researcher. In this case, the increase in knowledge, as a product of two-actor inquiry, is no longer an (additional) model or models of the actions of an (assumed) actor, but the increased ability of (second) actors to produce and maintain phenomena or behaviour modellable via such models.

Thus, allowing for the possibility to replace a model of an actor by the actor him or herself as 'researcher morphic', has considerable consequences. For knowledge now does not depend any more on a (necessarily) dominant context, a situation which may be called context-dependency free, or context-free. It depends on whether or not an actor wants to maintain the production of a model. Such a modelling choice therefore is more like an advice: do this or that to produce a '*model of*' which is suitable and advantageous for your particular context. As there is no order between first and second actors, such advice should result in an advantage or improvement, possibly different for both actors, due to different contexts. Such an advice thus comprises value elements (in a certain free sense, to be explained later), whereas a '*model of*' does not imply any choice or value for its user(s).

This change of conception of what it is that research produces, is important, especially in the framework of systems analysis. Systems analysis is a group of ideas to deal with practical problems, via research. Unfortunately it turns out, it is generally based on the one-actor model, and its application creates difficulties. Quite often systems analysis produces solutions with PS-type problems.[8] The above analysis may provide an explanation for this occurrence. It also suggests an (at least partial) redesign of systems analysis. Some elements of design in the two-actor model will now be discussed.

ELEMENTS IN THE TWO-ACTOR MODEL FOR INQUIRY

It is important to keep in mind the distinction between an *actor* 'in general' and a 'research actor'. The latter is an actor with specific constraints on possible aims and actions. One of these constraints consists of letting the product of action have a certain generalizability or quality of endurance over social, spatial and/or temporal relocation. An actor, in a more general sense, is someone who constructs an action as a certain order over time, and may reconstruct it using his or her other actions, both as material for, and as source of ordering principles. Such actions may be those of other actors, or of other research-actors.

The importance of this distinction lies in the fact that in the two-actor model, like in the one-actor model, only one actor needs to be a researcher. The point however is that other actors *may* become 'second' research-actors, as soon as they start to use the results and/or methods of the researcher actor and have to decide whether or not to co-operate in producing certain dominant contexts. However, the 'first' researcher cannot know *when* this happens. When the case obtains, we are dealing with the two-actor model for inquiry.

This distinction may make the following points easier to follow.

A. To whom, for whom, from where:

In actual two-actor research one should distinguish between the goalgroup, the addressgroup, and the sourcegroup. The *goalgroup* comprises those people or groups of people who should profit from the work of the research-actor. The *addressgroup* is composed of those who should act on the basis of the results of such work (and usually may react like the 'second' researcher). The *sourcegroup* consists of those people, or even of those solutions realized with people in the past, that may help in providing substance to the work of the research-actor.

In the last paragraph it was found that product from the two-actor model has the property of providing advice to the actors, of how to start producing certain '*models of*' from which they can obtain advantages—in terms of their own contexts. Here the various elements can be identified: someone is sent advice to do something (addressgroup), so there may be advantages for various groups of people in different contexts (goalgroup), and the advice is forged from contexts and actions provided by the actors (sourcegroup). In principle, these parts may coincide, but they need not: the researcher may first produce some advice, then check to whom it should be addressed, given that certain results are best for some specific people. The two-actor model thus does not imply that all actors are researchers, but that results should be checked as if one could have at least two researchers (even though one does not know whether this actually *applies*). As the one-actor model, the two-actor model is a quality control model.

As stated, the various groups may coincide, providing some special cases. If we take the goalgroup as 'all people', and the addressgroup equal to the goalgroup, results of research will satisfy the usual definition of knowledge: useful for all types of applications by anybody, useable by everybody (independent of society location, time-period). Sourcegroups may be statistical samples. Then such knowledge is in fact knowledge from the one-actor model, which turns out to be a special case of knowledge from the two-actor model. This is as it should be. One may derive the one-actor model from the two-actor model, by replacing one actor by a model of that actor, or more precisely, by replacing the addressee by his or her model. The two-actor model was derived the other way round.

In practice, the selection of the groups is a clear, though complex process. The goalgroup usually will be decided upon in discusions—political or economic (i.e. via who is providing financial support). The addressgroup will be selected by trial and error (in simple or sophisticated form): who will respond to advice in order that things start to happen to the advantage of the goalgroup? Note that Macchiavelli selected 'The Prince' as his addressgroup, for the benefit of the population (goalgroup). Nowadays, addressgroups usually are less restrictedly defined. Most times, the sourcegroup can be selected on a next-event basis: whenever more material of information is needed, select a source element that may provide the most of what is new, given what is found and used in preceding steps.

B. What is the result of two-actor inquiry?

Earlier it was indicated that when a dominant context is produced (temporally), one may find a model 'of' the resulting behaviour. Such a model of course depends on the process of producing the dominant context which will be called a '*model for*'.[7] Therefore, in the two-actor case, if one finds a '*model of*', it should be accompanied by a '*model for*'. Of course, this argument holds only when the position of one of the actors is taken: he or she is modelling what the other produces. However, such a position-selection was assumed to be impossible. Hence we actually can no longer know what is the '*model of*' and what is the '*model for*'. For the one actor the modelling produces a '*model of*', produced via the '*model for*' by the other. For the other, the '*model for*' must actually be the '*model of*' what the first is doing, i.e. in his or her (assumed) dominant context. And the '*model of*' must seem the '*model for*' which produces the first actor's results. Hence the conclusion holds that in two-actor inquiry there is a strong relation between '*models of*' and '*models for*'. However, their relative position is undecidable.

We may contrast this conclusion with the situation in one-actor inquiry. Here the '*model for*' selection of the set of phenomena to be modeled ('*of*') is usually called theory; and the phenomena, data. The relation between them is also taken as complex (at least in more modern versions of the one-actor model[9]). The data appear 'theory laden', i.e. contaminated in that they are, generally, assumed to have at least two conflicting roles: to infer the theory and to check the actor-independence. Even so, the relative roles of data and theory (in a dominant context) are taken, at all points in time, as strictly decidable and in fact hierarchical. This is clearly quite different from the situation in the two-actor type inquiry.

A non-hierarchical relation between '*models for*' and '*models of*' may be made somewhat more concrete as follows. Again, let us take the case of puerperal fever.[4] Reference to '*a model of*' turned out indications of some contagious substance (organism) transmitted by the doctors, presumably derived from the theory ('*model for*') that there was some (dangerous) connection between the doctors' two activities with pregnant women and with corpses. Later on, '*model for*' was developed to improve the situation: disconnect the two activities, e.g. via hand washing. This example was introduced as a clear example of a one-actor type of inquiry. However, suppose the two-actor model were to be used. Then, possibly, one would have found both the contagious organism, and the '*model for*', but one would not have presumed to know which was dominant. Was the advice 'to disconnect' a result of finding the organism, or was the organism 'produced' by the decision to disconnect?

Most people will automatically assume that the latter is not possible. And maybe they are correct. However, we may not be so certain when dealing with social systems (and possibly not even in some research situations in physics).[10] Examine the example of the noisy neighbour, was not his or her noise already a reaction ('*model for*') against paper-thin walls, produced therefore by installing such a de-coupling between neighbours—instead of the reason ('*model of*') for installing (more effective) types of de-coupling?

For two reasons it seems important, especially in such situations, to be aware of this non-hierarchical nature of '*models for*' and '*models of*'.

A. In the example of the noisy neighbour, if one were to assume instead a hierarchical relation, one might be led to think that the best thing to do would be to isolate him/her even more. This action may produce a desire in the noisy neighbour to be even more noisy—the relation would have been hierarchical, the other way. Thus, in this case, the assumption of hierarchy turns out to be the 'culprit' for the appearance of PS-type problems.

B. If one were not to assume such a hierarchical relation, then one might be in a position to prevent, or at least control, the appearance of such PS-type problems. One might then have started to change the relation between noise ('*model of*' or '*model for*') and partitioning via walls ('*model for*', or '*model of*') instead of react to only one type of model. An example of such a change might have been to make the effect of the noise on one actor clearer to the other, as a signal, and vice-versa. This might lead to an improvement for both. The point being that one would only be changing the relation between '*models for*' and'*models of*' (not knowing which is which), and thus prevent giving to any one of them PS-type problems (as a reaction to changing either the '*model of*' or '*the model for*', assuming one knows which is which).

As a general conclusion, we find that knowledge acquired via the two-actor model should be of the nature of advice, to change the relation between existing '*model for*' and '*models of*'. Such advice may therefore be: to reverse the order of the roles, to destroy any (presumed) knowledge about which model is a '*model for*' or a '*model of*'. It may also be a decoupling between '*models for*' and '*models of*', that would single out parts of existing models and interchange them to destroy any (presumed) hierarchical aspects.

Further elaboration of such types seems not necessary here. The main *aspect* has been demonstrated. Such knowledge should suffice: to make new 'changed' '*models of*' and '*for*' out of existing old ones.

C. What is accumulated when knowledge is acquired via the two-actor model? In the one-actor model such an accumulation clearly will consist of an increase of available '*models of*'. These would allow some durable qualities, that would allow to 'stack' them. But advice is not durable: one would need to have addressees able to use part of the '*models for*', to produce the '*models of*', as desired solutions. Such addresses might change over time and no longer know how to use the advice. Thus, the durability of advice must be of a different type. Advice itself does not accumulate. But what it brings about may be maintainable and hence enduring. Accumulation therefore will appear in the actual world of application, and not in the world of description of advice: various actors – both goalgroups and addresses – may change and improve what they can do together, to mutual advantage (if the two-actor model is followed). A general test of collective competence: the ability to work together, to the benefit of all (without benefit being necessarily similar in the various contexts of the actors).

D. What kind of testing is required to assure, as far as possible,...., that the acquired knowledge has the desired properties (non-reducibility to one-actor knowledge: non-hierarchy of '*models for*' and '*of*' action). The simplest answer here seems the one that is similar to the test procedure in the one-actor model. There, the test is to let the results of one researcher be checked by a second, etc. to ensure that what results can be accepted by all (from this result concepts like validity and reliability). The same type of procedure in the two-actor model for inquiry will be as follows: start with one research-actor, in a certain situation to be improved. Try to elicit some advice for improvement for him- or herself. Test with another actor, will such advice introduce PS-type problems? If so, change the advice such that PS-type problems are prevented for both actors. Consult a third actor, with the same question, etc.[11] Just like in the one-actor case, finding a first result (i.e. either a '*model of*', or an advice) may take some time (as there are various elicitation procedures[12]). However, the point should be clear, one may check by testing against other actor's contexts, to try to realize advice on a smaller scale, improve from there, etc. In these cases, one is testing by producing and changing contexts. One is not attempting to obtain a consensus.

CONCLUDING REMARKS

The main problem elucidated in this paper is to attempt to trace the causes for the appearance of so many post solution problems, especially in (seemingly) complex and sophisticated societies. Post solution problems imply synchronous problems, synchronous to the introduction of a solution, not the well-known ones that appear to spur researchers on to ever higher heights of insight. Much more can be said about such problems than what was presented here. However, I especially wanted to present some of my thoughts, first on how to interpret the occurrence of such problems, and second on how to proceed – at least partially – to prevent them. Summarizing my answers, my interpretation is that PS-type problems often stem from the process of finding solutions itself, a process which may need to be changed from a one-actor type of inquiring model to a two-actor type. Some specifications have been presented, and more can be found in the writing of many other authors (see also in this volume).

There are some more general aspects which I want to point out in the form of concluding remarks.

The first of these is that, what we are treating, concerns an old problem: how to relate many local worlds – which to some extent are autonomous (i.e. have local laws), and also seem to develop via each other. One approach is to find out what is common to all: what general laws and regularities apply. This has been a very frequent approach in the social sciences. Another has been to assume some specific commonality, like absolute time, the same in all worlds – as has been done in classical mechanics. I will not go further into the variety of such approaches. My main objective in mentioning them is to point out that indeed there is a problem with those kinds of inferences.

In the two-actor approach presented above, it was not suggested that one should look for what is the common law in various (partially) autonomous worlds. Rather, one should approach them by providing opportunities to develop, thus to build a record of interaction, and hence production. Such a production will indicate what is specific to such local worlds or actors. Therefore, the test of new knowledge about what holds such local worlds together is not that production can be predicted, but that new (unpredictable) productions are made possible and appear.

An important aspect is that such new knowledge is actually *about* the interaction of the local worlds, and is concerned therefore with what may be called *meta-worlds* to the local worlds. However, these are produced via the actor-researchers, with a non-hierarchical relation between '*models for*' (producing the meta-worlds) and '*models of*' (the worlds). One should not take the word 'meta-' to mean high. These meta-worlds are 'extra', but, generally, in terms of hierarchy they are undistinguishable from the local worlds. In essence, this means that accumulating knowledge will have the local worlds interact more, but will not imply more hierarchically complex organizations supporting such interactions.

To make the latter points more specific: what is necessary to prevent PS-type problems is to build '*models of*' the (organizational) meta-worlds, related to the '*models for*' of the actors- *next to* '*models of*' the actors for the meta-worlds. One should not only build regularity oriented imitations of actions[13] but rather more user-oriented action-supports. The former are seen as providing problems of design, the latter may be termed problems of 'meta-design'.

A second important general aspect is that in the two-actor approach no absolute time standard is required. Each actor may try to speed up what he or she is doing, by using more extensively the interaction with other actions and actors.

'Speeding up' in the sense that changes in an action can be controlled more easily. Thus there appear to be only relative time-speed changes, which relate various local worlds as a form of communication between the worlds.

A third question that may arise is whether there may be also three-actor models for inquiry, four-actor, etc. I have not gone into this question deeply. A simple answer may be that a two-actor model is satisfactory in two ways: firstly, it demonstrates already that the one-actor model need not be always used. Second, the non-hierarchical relation between '*models for*' and '*models of*' already implies the important fact, that any model of a third research-actor should be non-hierarchically related to those of the first two, etc. Thus the implication is that there is no common standard scale for any aspect (neither time, nor space, locality, structure, etc.). These properties seem to indicate that the two-actor model may already represent what we need, i.e. a strategy to prevent PS-type problems. However, it may be that in the future some multi-actor models will become useful, e.g. with the possibility of (temporary) coalitions between research-actors.

Finally, I want to emphasize that the above should make explicit that context-dependency can be handled as carefully and precisely as context-independency, an assumption which has always been assumed to be necessary to make 'hard' science. Hopefully, those who have been working on similar lines can gradually band together to actually make the distinction between hard and soft science obsolete. Meta-design need not be less precise or harder research than design.

REFERENCES

1. R. A. Nisbet, *Social Change and History. Aspects of the Western Theory of Development* (Oxford University Press, London, 1969).
2. C. Rubenstein, 'Vacations', *Psych. Today* **13**, 12, 62–77 (1980).
3. A. Tversky, D. Kahneman, 'Judgement under uncertainty: heuristics and biases', *Science* **185**, 1124–1137 (1974).
4. W. J. B. Beveridge, *The Art of Scientific Investigation* (Heinemann, London, 1950).
5. F. Benseler, P. M. Hejl and W. K. Köch, (eds.), *Autopoiesis, Communication and Society* (Campus, Frankfurt, 1980).
6. M. Hesse, *Revolutions and Reconstructions in the Philosophy of Science* (Harvester Press, London, 1980). W. Stegmüller, *Probleme und Resultate der Wissenschaftstheorie und Analytische Philosophie* (Springer Verlag, Berlin, 1973).
7. G. de Zeeuw, *Teaching modelling for change* (Published by Feoll, editor M. Lansky, 1983). G. de Zeeuw, *Modeldenken in de Psychologie* Diss. (Amsterdam, 1974).
8. G. de Zeeuw, 'Analysis of systems support. Some ideas on improving systems analysis', In: R. Tomlinson (ed.) IIASA publication, 1983 & E. S. Quade and H. J. Miser, (eds.) *Handbook of Systems Analysis* (Publication Draft IIASA, 1981).
9. C. W. Churchamn, *The Design of Inquiring Systems* (Basic Books, New York, 1971).
10. G. Zukav, *The Dancing Wu Li Masters* (Bantam Books, Toronto, 1979).
11. G. de Zeeuw, 'Is there really something better?' *SCIMA, J. Management and Cybernetics* **10**, 2–3 (1981–1982).
12. P. Groen, A. Kersten and G. de Zeeuw, *Beter Sociaal Weranderen* (Coutinho, Muiderberg, 1980).
13. G. de Zeeuw, 'Speeding up improvement', In: G. Lasker (ed.), *The Quality of Life and How to Improve it* (Pergamon Press, New York, 1981).

16. VERY RARELY STABLE

Ranulph Glanville

SUMMARY

In this paper,* I look at the architect's drawing as an example of a model which is dynamic: a model 'for', rather than a 'model of', in de Zeeuw's sense.[1] I explain how, using drawings, an architect can have a continuing conversation with himself, allowing him, thus, to develop his ideas in an essentially unpredictable manner. This is proposed as an alternative manner by which to think about and solve problems: a manner, nevertheless, entirely justifiable.

INTRODUCTION

The ideas of this paper have been elaborated elsewhere. I am not convinced that I have much to add, at a general level, to the position I had already published.[2] However, I am familiar with the mysterious world of architect's drawings which I promised to expose. It does seem relevant, for this book, to add a generalising framework that will articulate some of the more theoretical points, acting both as a bridge between my discussion of architect's drawings and my theoretical position.

To summarise, I believe that methodology, in itself, is a *meta field:* methodology is the method of method, ie. *meta method.* It is often treated, however, as if it were just another method. But method is, in itself (if we believe in the method of etymology) also a meta field: it is derived from the Greek *meta* and *hodos* (a way). So here we are in a whole pickle of metas, a real Gödelian mess, where methodology (let alone meta methodology) is already on a third level. Now, I happen not to believe much, any longer, in levels, except as convenient psychological constructs, for a number of reasons which I can adumbrate thus: boundaries cannot be ordinary lines, but must have a twist in them. The form of distinction is not a circle but a Möbius Strip.[3]

This leads me into a fascination with the collapsing of levels into each other,[4] and with circularities. I no longer consider circularity to be vicious (as a trained architect I doubt that I ever did, in a way which I hope this paper will make clear). On the contrary, I revel in it.

Consider, for example, the notion of stability. We have, normally, considered stability to be almost object like: a non-dynamic situation of the immutability of the state of affairs. But I believe we can no longer consider this to be so, in the light of recent cybernetic analyses of control, and notions of the cybernetics of cybernetics

* This paper is dedicated to my long-suffering students and teaching colleagues wherever they may now be. The title is taken from a sketch 'The Bruce's Philosopher's Song', by Monty Python's Flying Circus team. I should like to thank the Dutch Systems Group and specifically the editors of *Systemica* for allowing this expanded and English version of the paper to appear here.

(or the method of method, of course).[5] Stability, in this model, consists not in the controlled thing, but in the continuing interaction between the nominally controlled and nominally controlling. Thus, stability is dynamic. Method operates through action; it is not passive. The controlled thing may therefore appear externally unstable because the stable entity is the linked controlled/controlling pair. The communication between these two is where stability may lie. The controlled event (e.g. a conversation),[6] may thus appear, from the outside, to be *very rarely stable*. This is what the architect's drawing can help us understand.

In this paper, I explain how architects (or, more exactly, architecture students and teachers) use drawings to explore and develop ideas. This is a reflexive process. While the drawing is a model of an idea, it also affects that idea, allowing both development and greater definition. The idea changes, the circularity is complete, there is a new drawing. The drawing is modified. It is *very rarely stable*. On a couple of occasions I have used this (externally apparent) instability to actually and consciously generate designs.[7]

This process of enrichment, of the continuous collapsing of more and more detail into the idea,[8] is a reflection of my concept of a model in which model strings can be compressed, so that the next model actually appears to be a richer version of a 'previous' one, but not in fact a 'level down'. And the circularity I am claiming, is seen as a reverse algorithm, a process of anti-modelling, which allows the return to the fixed or transformable original idea, without improverishment.[9]

Before you read what I have to say about the mysterious world of the architect's drawing, I should add a couple of points. There is a 'proper' form for a scientific paper. Obviously, I have chosen not to follow it here. A new view of methodology and of models may require a new form of presentation. I try, when I write, to find a means of expression appropriate to – sympathetic to – what I am trying to say. Sometimes a less 'scientific' format seems appropriate.

There is a concept of a model which, in contrast to what we normally think about models, *is* very rarely stable. This is a concept which I use all the time in my normal everyday job – teaching architecture – and is embodied in the architect's drawing: not the sort made with precision drawing implements, but the sort sketched on detail paper or the back of an envelope, with soft pencil, and endlessly drawn over and redrawn. The process to which I am referring is much more general: it is, in fact, at the very root of the making and teaching of creativity – and I believe creativity can be taught.

Normally, a model is thought of as the simplified version of an already existing reality,[10] in which the salient qualities or characteristics of that reality are isolated. The form and pathways of their interconnectedness are established, together with the way in which these connections work – e.g. as functions or relationships which modify the behaviour of one (or more) characteristic(s), as they become another. Frequently, such models involve the assertion of theories, often arbitrarily postulated, black box like assumptions (e.g. the following relationship seems to describe a number of existing states, and is therefore assumed to be the correct relationship: that is, the one existing in nature'). Models, as generally thought of, can probably be divided into three types: those intended to describe, those intended to predict, and those intended to control. Predictive models are, in fact, used to control the future, as my next example illustrates. When NASA sent their astronauts to the moon, they used a descriptive model (and all of science may be seen as a descriptive model) – Newtonian Mechanics – to describe the salient qualities of local celestial motions, gravitational effects and the performance of their rocketry. (Einstein's Theory was far too sophisticated for NASA's purposes). But they used their descriptive model, as indeed science demands, predictively: that is, they used the model to control the future, to get the astronauts from the earth to the moon (at some time later) and then back again (later still). Indeed this sort of application of models can also be used to control in another, and quite different way – to make reality conform to the descriptive model itself. A very good example is Keynesian economics, where the only reality permitted or accepted is that described by Keynes (viz. the law of supply and demand), and where reality is adjusted until it fits the model. Marxist economics is, perhaps, even more like this, as exemplified by the overt (as opposed to the normal, but covert) rewriting of history to fit the model.

An example of the control model, is the railway signalling and switching model. A signalman, by controlling the simple states of the model – a set of lights etc. – does not *physically*, stop trains (there is no superman out there to physically arrest a train in motion): rather, he adjusts signals that instruct the train driver what to do, and hopes the driver will do it.[11] There isn't much prediction here, and the description is pretty simple and not very rich.

Such models are essentially stable. They don't change anything but the values of their qualities, nor do they in themselves change our ideas, understandings or intentions, although they may be used as the basis for new discoveries and, (especially the descriptive model), for the creation of a new descriptive (and stable) model[12], as in the change from Newton's to Einstein's view of the Universe.

Such models also involve an element of intention or purposiveness. The salient characteristics are isolated according to what we want to do – our interest in making the model. Equally, when we have various possible models, we tend to choose the one that most suits our purposes or agrees with our beliefs.[13] (The computational power needed to use Einstein's rather than Newton's model, made the choice of the former impractical).

However, there are models that are very different in their performance, which may involve no simplification (there may be very little simplification indeed, or they may even add information).

Such models can be placed under the general heading of 'representational models'. We normally consider many forms of verbal expression to be representations of some experience or event. Most of journalism, for instance, is of this type. So are the normal architect's drawings (as seen by the general public) of a building – although it should be noted that there is a difference here between two types of such drawing: the survey drawing which is a representational model of some reality (descriptive model) and the intentional representation of some future reality-to-be (a building to be constructed – rather like a moon landing). However, I am also interested in 'presentation drawings'. These are the (architect's) models the public see. While they are a very minor part of what the architect (among others) does, and although they may serve as a means of checking what should happen and one's understanding of it, they have little to do with the way in which architects develop their thoughts.

In the common understanding, words represent thoughts (of things).[14] But we are familiar with the way in which, when we are telling a true story, the words may take over and we suddenly find ourselves inventing new details (this can even happen with judicial evidence). A model is a means of exploration and development. Of course, there are other ways to design (e.g. computer aided/generated design). Let us now explore how architecture students produce drawings for an audience of teachers.

To set the scene: a room with perhaps 40 students in it, all seated at draughting boards: a small group of staff wandering around, looking over shoulders, occasionally sitting down with students, talking and drawing with them; a project, to design a house for a park-keeper on a site in Regent's Park, London – or, something probably rather more abstract, such as a wall surrounding an urban hole, waiting to be filled.[15]

How do we proceed? The problem is the following: the student may have the beginnings of some idea, but in order to explore it, he must represent it (make a model of it), usually initially with paper and pencil (architects are told to 'think with their pencils'). Before the student has thus articulated the idea, the teacher, of course, has no access to it and even when the idea is 'represented' he has no way of accessing the idea as it exists in the student's mind – only his (the teacher's) idea (of that idea). This is a very real problem in all such paradigms of representational models: one ends up with a regress of (my) idea of (your) idea of (my) idea of (your) idea . . . etc. which can only really be broken by asking 'if I put it this way, does it make sense to you?'[16] So the student maybe starts with the inkling of an idea somehow half-formed, and sits down to represent it in a drawing with a few words. This information is usually quite vague and the idea is very rarely complete. In representing it, the student is exploring, testing and hoping to develop the idea. The drawing is a sketch and through this sketch he can exteriorise the idea, see its strengths and weaknesses. He may, indeed usually will, play with several such ideas – often interjecting them in what might be seen as the logical and sequential growth and elaboration of an earlier idea which may be discarded. In fact, what he is doing, is having a conversation with himself, by sketching ideas on paper.[17]

The mere fact of expressing ideas changes them: Not only because they are probably ill-formed or because they are used for further development with all sorts of modifications, but because something does not work or because there are more ideas to be incorporated or more requirements to be resolved.

I cannot fully explain these conditions here, but basically, the following takes place. In the act of representation, two things (e.g. idea and drawing) are brought into a sense of identity. But they cannot be exactly identical since they are separate and therefore must show some differences.[18] There is, therefore, always distortion, or change: the two elements in the representation are not the same, and to look at the drawing and use it, itself, as part of another representation (as the teacher must do and the student, returning to a sketch after a little time will do) means there must be change. In this way, the mere act of representation, seen in these terms, strongly implies a conversation, and the main purpose of representation, must be seen as implying just such a conversation. The use an architect, student and teacher make of such drawings is conversational – in the student's case, a conversation with himself and, less importantly, with his teacher.

All this confirms the point I made earlier about the teacher's necessary initial ignorance (teachers who have denied this, teach by rote, and, brainwash, even more than our schooling already has, the creativity out of students). I do not, here, refer to a teacher's professional expertise, or to experience. Both are essential and helpful to students. Rather, I refer to the initiation and development of ideas. Quite simply, the student's idea is his and not the teacher's, and the teacher has absolutely no access to it (barring telepathy) until the student has begun to represent the idea, at which point the teacher may form his own idea of the (student's) idea and join in the conversation. And through these conversations[19] – the student's own one and his one with the teacher – the transition from nothing to a workable idea is, in a faltering way, achieved: something comes out of nothing and that is creativity: creativity is teachable. (I have described design as being like wandering, never being quite sure where you are going but suddenly realising you are there!)[20]

Of course, these drawings – representational models – are, necessarily, *very rarely stable:* they are meant to be. They are 'models for' – models that enable the achievement of some final and initially only hazily clear end – not 'models of', which occur when you have achieved your previously unspecified end (which becomes an end by recognition and is justified by post-rationalised explanation).[21]

I am tempted, here, to put forward this idea of the self-conversation through representation, and of teacher ignorance being even greater than that of the student, as necessary (though not sufficient) conditions for the development of creativity. Indeed, I would assert that the acquisition of the skill of assumed ignorance (what I call the 'Principle of Ignorance'[22]) is essential to creativity. After all, we have learned all we know from a condition of extreme ignorance in infancy,[23] and we can never see things anew if we consider all our opinions fixed or as if we already know the answer. I had some idea of what I wanted to say when I began writing this paper, but it neither resembles what I said when I first discussed it with the editor, nor did I know how to say it when I began, nor is the way I first wrote it, how I see it now.

I believe we all can (and probably do) work like this, and the substitution of methods that prevent us from doing so is anti-human. It is the method of design and is a valid general method.[24] If designers don't succeed, it is because we expect too much of an indeterminate and open process we understand too little, and expect to interfere with and eventually replace it with those more 'quantitative' and 'exact' methods to which we are so partial and which have never yet produced design of any quality. In our desperate search for the stable, we are besotted with 'models of' instead of 'models for'. Yet much of what we must do is, in the doing, *very rarely stable*.

REFERENCES

1. The concept 'model of'/model for' are developed by G. de Zeeuw *Modelenken in Psychologie* PhD. Dissertation (University of Amsterdam, 1974). He encouraged me to

write my ideas for the present paper. G. de Zeeuw's ideas are presented in this book in chapter 15.

2. This research is most recently elaborated, much along the lines indicated here but in much more detail, in my paper 'Method in Methodology' in which I discuss at length the meta concept and the consequent need for action (in L. Troncale, (ed.) *A General Survey of Systems Methodology*, Vol. 1. (Society for General Systems Research, Louisville, KY, 1982). It refers to my paper 'The Model's Dimension: a Form for Argument' which I introduce in a later note.
3. See R. Glanville and F. Varela, 'Your Inside is Out and Your Outside is In' in G. Lasker, (ed.) *Applied Systems & Cybernetics* Vol. 2, (Pergamon, Oxford 1981), a development of my 'Beyond the Boundaries', in R. Ericson, (ed.) *Proceedings of the Society for General Systems Research*, Silver Jubilee Conference (Springer, London, 1979), which discuss George Spencer Brown's Logic of Distinctions, *The Laws of Form*, (George Allen & Unwin, London, 1969) in terms of the reputed distinction between mark and value, or the distinction of being Möbius-strip-like, rather than circular. In terms of levels, I develop this idea in my 'Levels & Boundaries of Problems' (see a later note).
4. G. Pask ('Lp – a Proto-Language') Laboratory memo, 1979, refers to this sort of process as 'Condensation'. I have looked at this process in terms of the reputed levels of Cybernetics – e.g. Second-Order Cybernetics – in my paper 'What a Waste', submitted for publication to *Cybernetics Forum*.
5. The American Society for Cybernetics Conference in Washington D.C. (1981) concentrated on this topic – tape transcripts are available from the Society. My 'Inside Every White Box There are Two Black Boxes Trying to Get Out' explores, explicitly, such notions of control (*Behavioural Science* **12**, No. 1, 1982).Heinz von Foerster's *The Cybernetics of Cybernetics* (BLC document 73.38 University of Illinois, Urbana, 1979), is the basic reference for these issues.
6. I am not, here referring to a technical 'conversation' in Pask's sense. (see later note.)
7. This is described in a recent document of mine, 'Motivation', available on request.
8. I have discussed this in terms of spatial perception in 'Amazing Space! For the Architectural Stimulus-Response Rat?' (*Architectural Association Quarterly* **9**, Nos 2–3, 1977).
9. I have written about a highly stripped down version of some of these sorts of modelling procedures in three papers. The basic theory is expounded in my 'The Model's Dimensions: a Form for Argument'. (*Int. J. Man-machine Studies* **13**, 1980). I have applied it to psychological processes in 'All Thoughts of Things' in *Progress in Cybernetics & Systems Research*, Vol. 9, (Hemisphere, Washington D.C., 1982) and, in a computing application with Peter Jackson ('The Theory of Model Dimensions Applied to Relational Data Bases', submitted to *Int. J. Man-Machine Studies*), in which an extremely simple problem solving program is put forward. The signalling example (see note 11, *Op. cit*) is in the first mentioned paper.
10. I use the term 'reality' here as a convenience. I do not believe in conventional notions of reality as being 'out there', or 'what we connect with' any more (as will be apparent from my later assertions about intentionality in modelling) than I believe in the naive view of 'the objective'. As far as I am concerned, reality is something we invent – possibly by, but in no way without, our interaction with what we may postulate as 'it'. This view is no longer so exceptional, and may be called 'Construction'. Of the many people who have recently espoused such a view, perhaps Piaget's *The Child's Construction of Reality*, (Basic Books, New York, NY 1955) and von Foester's (*On Constructing a Reality Research*, (Downden Hutchinson & Ross, Stoudsburg, PA, 1972) are the most notable.
11. See Note 9, *Op. Cit.*
12. This idea of a complete shift in scientific thinking has been explained by a number of philosophers of science, but notably by the physicist Thomas Kuhn in *The Structure of Scientific Revolutions* (Chicago University Press, Chicago, 1970) in which his idea of a paradigm shift is presented. Actually, I prefer Imre Lakatos' concept as expounded in

'Falsification and the Methodology of Scientific Research Programmes', in I. Lakatos, and A. Musgrave (eds.) (Cambridge University Press, Cambridge, England, 1970).

13. Another, and more graphic, example may be found in map making, which I have written about in 'Mapping Realities' *Architectural Association Quarterly* **12**, No. 4 (1980).
14. I would refer, here, again, to my paper 'All Thoughts of Things', *Op. Cit.*, and also to the middle paper of my trilogy 'Reasons to be Cheerful' (Parts 1, 2 and 3), presented at the Sixth Annual European Meeting on Cybernetics and Systems Research in Vienna, (1982). The latter is strongly influenced by the drawings of my many students which I observe from afar.
15. There are, of course, normally specifiable functional requirements – although these may often have to be discovered as well. What I am getting at here is architectural and social quality.
16. There is a theory that deals with this problem called 'Conversation Theory' for example G. Pask *Conversation, Cognition & Learning* (Elsevier, New York, 1976). In this theory, the relative exploration of one's own concepts (ideas) vis-a-vis those of a conversational partner are thoroughtly analysed, and a whole calculus for dealing with them is developed.
17. There are naturally many other ways also used to develop such ideas, for example, calculation. But these should not be allowed to deny or exclude what I am discussing here.
18. I discuss this at length in the part 2 of my paper 'Reasons to be Cheerful' (Parts 1, 2 and 3), presented at the Sixth Annual European Meeting of Cybernetics and Systems (Vienna, 1982) – the source of which is my doctoral thesis *The Object of Objects, The Point of Points – or, Something About Things*' (Brunel University, Uxbridge, 1975). A similar point is made in terms of physics, in my 'The Same is Different' In: M. Zeleny, (ed.) *Autopoiesis* (Elsevier, New York, 1981).
19. As an extension of the informal conversations as we probably all understand them, see G. Pask (*Op. Cit.*).
20. I discuss this process in my paper 'The Architecture of the Computable' *Design Studies* **1**, No. 4 (1980), which was, ironically, redesigned by the editors of the journal before publication.
21. See Note 1, *Op. Cit.*
22. See, for instance, my paper 'Levels & Boundaries of Problems' delivered to the Conference on Problems of Levels & Boundaries, Amsterdam, 1981. This paper postulated that levels and boundaries were purely cognitive artefacts, which generate problems. There are no problems other than those we are prepared to construct! In this paper, I argue for ignorance as a basis for learning and creativity, a point frequently forgotten by teachers. Also my invited address 'A Cybernetic View of Education and a View of Cybernetic Education', given to the American Society for Cybernetics (A.S.C.) Conference in Columbus, Ohio (1982).
23. See Piaget and Constructivism (*Op. Cit.*) There is some confusion about intuitive behaviour, but I believe I have answered this doubt in a constructive manner in 'All Thoughts of Things' and 'The Architecture of the Computable' (*Op. Cit.*).
24. See 'Why Design Research', in R. Jacques and J. Powell, *Design Science Method* (Westbury House, Guildford, 1980) where I argue that science is a sub-branch of design and not vice versa. Therefore, it is wrong to judge design by scientific standards although alright, to judge science by design standards.

17. THE METAPHYSICS OF SOCIAL SYSTEM DESIGN*

Werner Ulrich

SUMMARY

Different systems paradigms imply different models of the architecture of complexity and thus different strategies for dealing with ill-structured problem situations. Herbert A. Simons's model of complexity, which underlies the currently prevailing systems paradigm of Systems Science as well as management and planning theory, is critically discussed with respect to its metaphysical and normative content. To this purpose, a fictitious 'debate' is held between H. Simon and C. West Churchman, whose philosophy of social system design relies on an alternative model of complexity. In this way, the reader is encouraged to reflect on his own metaphysical preconception of the social systems with which he is concerned, be it as a planner, administrator, manager, or simply as a citizen.

> The task of our generation, I have no doubt, is one of metaphysical reconstruction . . . We are suffering from a metaphysical disease, and the cure must therefore be metaphysical.
>
> E. F. Schumacher
>
> *Small is Beautiful* (1975)

THE MANAGEMENT OF COMPLEX SYSTEMS, AND THE METAPHYSICS OF SYSTEMS MANAGEMENT

At least since the Apollo Project of NASA, the systems concept has been associated by the public with the management of complex decision-making and planning processes. It has hardly been noticed, however, that the systems concept presupposes a metaphysical preconception of the nature of 'complexity'. Without some previous understanding of the nature of complex social reality, the systems concept cannot help us to comprehend and manage this reality.

The systems scientist, Herbert A. Simon, was awarded the 1978 Nobel prize for his contribution to the understanding and management of complex decision-making processes in social systems. Especially in his paper on 'The Architecture of Complexity', Simon[1] has given its most elegant formulation to the model of complexity that underlies contemporary management and planning thought. But the elegance of the formulation is not necessarily a guarantee of the validity of the content. Let us therefore take the first Nobel award for systems science as providing us with an

* An earlier, shorter version of this paper was published under the title 'The Metaphysics of Design: A Simon-Church 'Debate' in *Interfaces* **10,** No. 2, 35-40 (1980). The present version appeared in the author's book, *Critical Heuristics of Social Planning* (Bern, Switzerland: Paul Haupt, 1983).

opportunity to reflect on the metaphysical content of the prevailing systems concept. In what way is its underlying model of complexity a source of hidden value content and/or practical irrelevance of contemporary management and planning theories?

A fictitious debate* between Herbert A. Simon and another outstanding philosopher of systems management, C. West Churchman, will help us to reflect on the metaphysical content of Simon's systems concept. At the same time, this 'debate' may serve the reader to clarify some basic metaphysical presuppositions of his own conception of social systems design.

THE CONCEPT OF COMPLEXITY IN TWO ALTERNATIVE TRADITIONS OF THOUGHT

Before entering into the debate, let us briefly introduce the two debaters and their views of complexity. The two standpoints represented in our debate can be labelled by the titles to two of their works: Churchman's *The Design of Inquiring Systems*[2] and Simon's 'The Architecture of Complexity'.[3] Two labels also used by the authors are '(The Philosophy of) Social Systems Design'[4] and *The Sciences of the Artificial*.[5]

Simon[6] calls a system 'complex' when 'the whole is more than the sum of the parts, not in an ultimate, metaphysical sense, but in the important pragmatic sense that, given the properties of the parts and the laws of their interaction, it is not a trivial matter to infer the properties of the whole'. This 'pragmatic' definition is, however, ambiguous with respect to its metaphysical content. From what has been said above it is clear that Simon cannot avoid a metaphysical pre-judgement in applying his systems concept. A closer look reveals that Simon's assertions betray the metaphysical content of his seemingly metaphysics-free, if not antimetaphysical, systems concept. A good example is his reliance on 'in-principle reductionism'. 'In the face of complexity, an in-principle reductionist may be at the same time a pragmatic holist'.[7] It is obvious that even a pragmatically disguised reductionism will determine the way in which a planner tries to understand and master complex real-world problems.

By *complexity* we therefore have to understand not so much a 'given' attribute of a particular system but rather an individual strategy for dealing with ill-structured problems. That is to say, a model of complexity is nothing other than an inquirer's way of confronting a lack of knowledge — the lack of empirical or logical decidability that is characteristic of all metaphysical assumptions.

Such a concept of complexity is of course quite different from the definition offered by the *Oxford English Dictionary:*

Complex, a. (f. *com* — together + *plexus* — plaited . . .)

1. Consisting of or comprehending various parts united or connected together; formed by combination of different elements; composite, compound. Said of things, ideas, etc. (Opposed to *simple,* both here and in sense 2.)

* Both Professor Churchman and Professor Simon have kindly agreed to the publication of this 'debate'. As I worked with West Churchman from 1976 to 1980, it is clear that my conclusions (but not the debate itself) lean towards his position; I am all the more grateful to Professor Simon for accepting my account as a 'fair' description of his position and for permitting its publication.

2. *esp.* Consisting of parts or elements not simply co-ordinated, but some of them involved in various degrees of subordination; complicated, involved, intricate; not easily analysed or disentangled.

This definition neglects virtually all those aspects of complexity that are crucial to a planner. A heuristically useful concept of complexity ought to consider the root causes of lacking decidability, for example the inseparability of systems components for the purpose of planning (design-nonseparability); the uniqueness of systems (and hence their design-nonfeasibility); the intrinsic autonomy and self-motifivation of systems; the perspectival character of any standpoint from which the planner may apprehend a system (for example the dependency of any definition of a system on an interest): The *Oxford* definition is obviously based on the Aristotelian, analytic-reductionistic tradition of thought which is so familiar to Western thinking that it is identified with common sense, even though it depends on very questionable (because metaphysical) assumptions about complexity. Herbert Simon is an outstanding representative of this tradition, a tradition that has built its enormous success on Ceasar's old principle: *divide et impera !* — translated into the language of systems science: decompose and control!

C. West Churchman's philosophy of social systems design, on the other hand, is closer to the other great tradition of Greek philosophy, the nonobjectivist and antireductionistic tradition of thought of which Plato is representative. To avoid the derogative meaning with which the dominant Aristotelian tradition has associated the term 'Platonic', it may be advisable to speak of 'dialectical' or simply Non-Aristotelian rather than 'Platonic' thinking. This tradition has recently gained new attention because it promises more help in dealing with design-nonseparability, uniqueness, and other essential difficulties encountered by the social planner.

For the planner who understands himself as a systems designer, the two positions represent more or less opposite answers to crucial questions such as wholeness or decomposability of systems, the 'spirit' of social systems, the role granted to subjectivity in one's understanding of 'rational' design and so forth. Opposite need not imply incompatible; indeed the most significant difference between the two positions might be that the one (Simon) *in principle* denies their compatibility whereas the other (Churchman) does not. Churchman has argued that either position becomes absurd, if taken as absolute; both are needed for rational thinking about the reality of systems.[8]

The stage is set; enter the two debaters.

Position:	"Sciences of the Artificial"	"Philosophy of Social Systems Design"
Represented by:	H. A. Simon	C. W. Churchman
Subjectivity is	*avoided:* excluded from the definition of objectivity; objectivity has no way to deal with the spirit of social systems	*not avoided:* must be included in the definition of objectivity; the designer must consider the spirit of individuals and whole systems in his design
Complexity is	*hierarchical:* complex systems are built up from hierarchies of subsystems	*not hierarchical:* systems are complex because they cannot be described as hierarchies of components

Position: *Represented by:*	"Sciences of the Artificial" H. A. Simon	"Philosophy of Social Systems Design" C. W. Churchman
Wholeness of systems (holism) is	*denied except in a pragmatic sense:* the whole is not more than the sum of the parts, though its analysis is not trivial; regarding complex systems as wholes does not help explain them	*critically considered:* there are wholes that cannot be dealt with as sums of parts; regarding complex systems as wholes serves the critical purpose of reminding us of the limits of our understanding
Reducing complexity of systems to "simple" subsystems (reductionism) is	*a source of knowledge:* every system has components which are simpler systems; systems can be understood as black boxes, and the inquirer's task is to make them transparent (reductionistic systems approach)	*a source of irrelevance:* what really matters is systems that are individuals; the inquirer's problem is to account for the aspect of systems (antireductionistic systems approach)
The purposeful character of systems (teleology) is	*denied:* ascribing purposes to social systems is a source of deception (teleological fallacy)	*critically considered:* ascribing purposes to social systems serves a necessary critical purpose against hidden value assumptions (teleological imagery is needed in addition to causal-analytic terms)
The *decomposability principle* is	*accepted:* near-decomposability of complex systems (the decomposability principle for artificial systems is applied to social systems)	*rejected:* design-nonseparability of the components of complex systems (the design-separability of components of artificial systems does not hold for social systems)
The *description of systems* must	*focus on redundancy:* by making use of the world's redundancy, complex systems can be described simply (e.g., in terms of simple process descriptions standing for redundant state descriptions)	*focus on uniqueness:* every system has not only redundancy but also uniqueness; to the extent a system is unique, it is its own simplest description (description of uniqueness matters as much as description of redundancy)
The *crucial design task* to be solved is	*problem decomposition:* how to decompose a complex system into simple systems that are easy to be controlled? Or: how to design and control complex hierarchies? ("*Divide et impera*!" standpoint; control problem)	*problem identification:* how to describe unique systems, and especially: how to identify the whole system on the one hand, the smallest system (individual) on the other? ("*Ethics of whole systems*" standpoint; boundary problem)
The designer's *main tool* is	*objectivity:* semantic precision of concepts, model building, causal-analytic explanation: mathematical analysis empirical research, computer simulation, heuristic programming; scientific rigor "programmed decision making" (*Science of the Artificial*)[9]	*subjectivity:* reflection on the sources of knowledge and deception: social practice, community of inquirers; interest and commitment; a priori ideas (systems idea, moral idea, moral idea, guarantor idea); "process of unfolding" (*The Artificiality of Science*)[10]

Table I: A ficticious "debate" on the metaphysics of systems design.[11]

SOME IMPLICATIONS

From a design point of view, the major difference emerging from the debate concerns the authors' view of what constitutes the crucial lack of knowledge, i.e. the complexity of real-world problem solving and decision making. For Simon it is objective *explanation,* for he identifies the limits of rational design with those of rational (= objective) explanation. For Churchman it is the consideration of and *design for the inexplicable,* because design that does not consider the whole system invites deception.

The underlying motivation is sincere and valid in both cases, so long as no claim for generality is linked to it. Herbert Simon[12] (also Newell, Shaw and Simon[13]; Newell and Simon[14]) attempts to apply the objective approach of natural science to the domain of social inquiry and design. In Simon's terminology, all systems designed by man are called artificial, in contrast to nondesigned natural systems; hence the name 'Sciences of the Artificial'. The underlying assumption is that we cannot design subjective social systems. This is true from an objective point of view. It is however, irrelevant from a critical point of view: man cannot escape his freedom to design *for* the social systems of which he is an integral part. And it is a source of deception, for it fails to consider the fact that all design (or artifical systems) is design directed to some subjects to be served.

C. West Churchman's philosophy of social systems design,[15] on the other hand, is an attempt to overcome the gap between subject and object wherever it prevents the planner from dealing with the subjectivity of the systems for which he plans. Churchman tries to include the subjectivity of systems in his concept of objectivity rather than to avoid it.

To make it very clear: the question is not whether one can objectively explain and design social systems as whole, unique systems. The question is whether or not we want to design *for* them to become more whole, unique, self-motivated, and so on.

The first rather than the second question has been addressed by Simon's theory of complexity. That may explain, to a large extent, why it has become a major source of irrelevance — if not deception — in design *for* social systems: General Systems Theory, RAND-systems analysis, systems engineering, managerial cybernetics[26], artificial intelligence, heuristic programming, MIS design, Management Science/Operational Research, incrementalist planning approaches, behaviourist learning theories, administrative and policy science all suffer from a common *inability to deal critically with the very social reality which they ought to improve.* Langdon Winner[16] has traced a bit further the limited rationality of these systems-scientific approaches to social complexity. According to Winner, they rely on four typical assertions:

(1) 'The first prescription for (dealing with) complexity is always the same. Simplify, In modern practice, this usually means 'control as much as possible of that which is changing in the system'.[17]

(2) 'Technology caused the problems. Technology must solve them'.[18] The progress of computer *hardware* and software (which in fact is rather 'hard'!), together with systems analysis, is the supposed remedy for complexity.

(3) Another remedy has been found in the functional *redundancy* of complex systems which is taken for a guarantor of complexity control.[19] 'By implication this argument gives us one reason why failure to understand complex systems may not be

all that important as a practical matter. As long as the complication of an organisation includes sufficient redundancy, the organisation will be protected from serious errors and breakdowns. Complexity in this sense contains its own saving remnant.[20]

As a comment of this, it may suffice to point to the nuclear incidents at Three Mile Island near Harrisburg, Pennsylvania, in April, 1979 or to the disaster at Chernobyl, USSR in April 1986. In the design of nuclear power plants technical redundancy of emergency systems has increasingly been substituted for subjective and social rationality as a guarantor of complexity control.

(4) Finally, the idea of disjointed *incrementalism*[21] is often inherent in this position, although the notion as such is now in disrepute. 'All that a decision maker needs to understand is the specific context of the activities of the particular choice before him. Knowledge of the activities of the whole system is irrelevant except as it bears directly on the incremental step at hand'.[22]

Incrementalist thinking is presupposed in Simon's principle of 'near-decomposability' and many other concepts currently used in planning, from cybernetics to interest-group pluralism. The problem of incrementalism is clear: it simply avoids the moral and political question of how a better 'whole' will result from incremental improvement of small and separate subsystems; it leaves that question to the market as a surrogate planner — and to the interest groups that control it as surrogate clients.

The second of the above questions — of how we can design *for* social systems to become more whole, unique, etc. — is addressed by Churchman's theory of complexity. He associates the planner's unavoidable lack of knowledge with the unique and nondecomposable character of social systems. The crucial task then is not one of providing analytic tools such as social indicators, simulation models, etc., but rather dialectic tools to help the planner reflect on his designs, and enter into reasonable discourse with the affected. In his book, *The Systems Approach and Its Enemies,*[23] Churchman accordingly presents a dialectical approach to the unfolding of the unavoidable conflict between the systems planner who strives for systems rationality and the 'enemies' of such rationality who are concerned in their private, subjective rationality. The point is that no systems approach for handling real-world complexity can be rational unless it makes room for the (self) critical reflection of free *citizens;* for these alone know what the social reality to be improved is like. The Systems Approach may yet have a great future, if only it begins to understand that the 'enemies' are really its allies.[24,25]

REFERENCES

1. H. A. Simon, 'The Architecture of Complexity', *Proceedings of the American Philosophical Society* **106,** 467-82 (1962).
2. C. W. Churchman, *The Design of Inquiring Systems, Basic Concepts of Systems and Organisation* (Basic Books, New York 1971).
3. cf. reference 1.
4. C. W. Churchman, *The Systems Approach and its Enemies* (Basic Books, New York 1979).

5. H. A. Simon, *The Sciences of the Artificial* (MIT-Press, Cambridge, Mass. 1969).
6. H. A. Simon, 'The Architecture of Complexity', ref. 1, p. 468.
7. H. A. Simon, 'The Architecture of Complexity', ref. 1, p. 468.
8. Cf. C. W. Churchman, *Challenge to Reason* (McGraw-Hill, New York, 1968) Ch. 14.
9. Cf. Reference 5.
10. C. W. Churchman, 'The Artificiality of Science', a review of Herbert A. Simon's *The Sciences of the Artificial, Contemporary Psychology* **15**, 385-86 (1970).
11. Adapted from W. Ulrich, 'The Metaphysics of Design: A Simon-Churchman 'Debate', *Interfaces* **10**, No. 2,35-40 (1980).
12. H. A. Simon, *Models of Man* (Wiley, New York, 1957); H. A. Simon, 'The Architecture of Complexity'; H. A. Simon, *The Sciences of the Artificial Op. Cit.*
13. A. Newell, J. S. Shaw and H. A. Simon, 'Elements of a Theory of Human Problem Solving', *Psychological Review* **65**, 151-66 (1958); A. Newell, J. C. Shaw and H. A. Simon, 'Report on a General Problem-Solving Program for a Computer', In: *Information Processing, Proceedings of the International Conference on Information Processing* (UNESCO, Paris; and Butterworths, London 1960) p. 256-44.
14. A. Newell and H. A. Simon, *Human Problem Solving* (Prentice-Hall, Englewood Cliffs, N.J. 1972).
15. C. W. Churchman, *Challenge to Reason;* C. W. Churchman, *The Systems Approach* (Delacorte Press and Dell Publishing, New York, 1968, rev.ed. Dell Publishing, 1979); C. W. Churchman, *The Design of Inquiring Systems;* C. W. Churchman, *The Systems Approach and Its Enemies;* C. W. Churchman and W. Ulrich, 'The Status of the Systems Approach, Reply to Bryer' *Omega* **8** 277-80 (1980); C. W. Churchman, T. A. Cowan and W. Ulrich, 'The Systems Approach and Its Enemies: A Dialogue', *J. Enterprise Management* **3**, 200-202 (1981); C. W. Churchman, *Thought and Wisdom* (Intersystems Publications, Seaside, Ca., 1981); W. Ulrich, Review of Churchman's *The Design of Inquiring Systems, J. Operational Research Society,* **36**, 873-875 (1985).
16. L. Winner, 'Complexity and the Limits of Human Understanding' In: T. R. Laporte (ed.), *Organized Social Complexity* (Princeton University Press, Princeton, N.J. 1975) pp. 40-76.
17. L. Winner, 'Complexity and the Limits of Human Understanding', p. 65.
18. L. Winner, 'Complexity and the Limits of Human Understanding', p. 65.
19. Cf. M. Landau, 'Redundancy, Rationality and the Problem of Duplication and Overlap' *Public Administration Review,* **29**, 346-58 (1969).
20. L. Winner, 'Complexity and the Limits of Human Understanding' p. 69.
21. R. A. Dahl and Ch. E. Lindblom, *Politics, Economics and Welfare,* (Harper, New York, 1953, 2nd ed. with a new preface, 1976); Ch.E. Lindblom 'The Science of 'Muddling Through'' *Public Administration Review* **19**, 79-88 (1959) L. Braybrooke and Ch.E. Lindblom, *A Strategy of Decision* (Free Press, New York, 1963).

22. L. Winner, 'Complexity and the Limits of Human Understanding', p. 71.
23. C. W. Churchman, *The Systems Approach and Its Enemies, Op.Cit.*
24. For a comprehensive effort to elaborate the epistemological and heuristic foundations of a critical systems approach to planning, an effort inspired by Churchman's work, see W. Ulrich, *Critical Heuristics of Social Planning, A New Approach to Practical Philosophy* (Haupt, Bern. Switzerland 1983).
25. As a first introduction, see W. Ulrich, 'Critical Heuristics of Social Systems Design' *J. American Planning Association,* to appear; M. C. Jackson, Review of W. Ulrich's *Critical Heuristics of Social Planning, J. Operational Research Society,* **36**, 878-881 (1985).
26. W. Ulrich, 'A Critique of Pure Cybernetic Reason, The Chilean Experience With Cybernetics', *J. Applied Systems Analysis,* **8**, 33-59, (1981).

18. THE METASYSTEM, RATIONALITIES AND INFORMATION

John P. van Gigch

SUMMARY

The metasystem approach is used to formulate the distinction between data, information, and intelligence. These forms of communication are conceptualised in terms of four types of rationalities which must be present to constitute a message that can lead to a goal-oriented decision. The importance of these factors for the design of viable systems which can avoid failure is explained.

PURPOSE

This chapter is concerned with the metaorganisation of information. Its intent is:

1. To place the various forms of information in a metasystem framework. To this end, the metasystem approach and the metasystem paradigm are first explained.
2. To discuss the components of communication, data, information, and intelligence, in relation to this framework and
3. To infer the requirements which need to be placed on these components to avoid systems failures.

THE METASYSTEM PARADIGM

The metasystem paradigm is based on the model developed by Kickert[1] and extended in Kickert and van Gigch[2] where decision making is studied from a control system's viewpoint. The reader is referred to Chapter 3 of this book where the basic concepts are defined. There, we recall, the control system (C) is conceptualised as made up of a controller (CR), a controlled system (CS) and an environment (E), plus a metacontroller (MCR). A control system (C) and its metacontroller (MCR) define a level of recursion – a term used by Beer[3–5] to designate control systems at different levels of logic. Higher levels of recursion are defined as more metacontrollers (MCR', MCR''. . . etc.) are added. Control exercised by the controller (CR) and by the metacontroller (MCR) are called *object level control* and *metacontrol*, respectively. (See Figure 1.)

As a methodology by which decision making can be studied, the metasystem paradigm can be compared to the science and to the systems paradigms. In Chapter 2 of this text the strengths and weaknesses of these and related methodologies are discussed.[6]

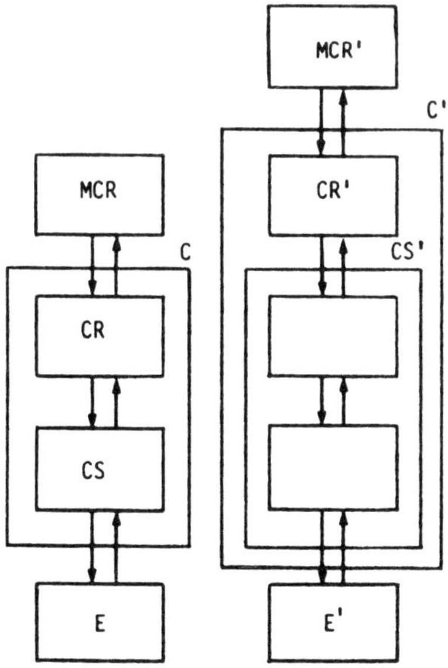

Fig. 1. Control systems C and C′ with their respective metacontrollers MCR and MCR′ define two levels of recursion.

RATIONALITIES

In order to describe the information flows which guide behaviour and decision making at each level of recursion, four different types of rationality are defined. *Rationality* is to be considered synonymous with reason, argument, motive, cause, or justification for a particular behaviour. Behaviour and decision making are considered *rational* if they are consistent and are justified by the precepts, norms and guidelines of the said rationality. Rationality is a term which was used originally by Simon.[7–9] We modify slightly the definition given by Simon to substantive and procedural rationality and add two other rationalities. We distinguish four rationalities as follows:

1. Structural rationality (R_s) guides the establishment of the structure of organizational decision making. Changes of structure occur during adaptive internal (IA) and adaptive external (EA) control at the object level and during similar modes (IMA and EMA) at the metacontrol level. As stated by De Leeuw[10,11] and Kickert,[1] structural rationality is always involved with the organisation of decision making where *structure* of a system is defined by the set of relationships between subsystems (groups, departments: WHO decides?) between aspect systems (the issues, topics about which decisions are taken: WHAT is decided and HOW?) and phase systems (the phases: WHEN are decisions taken?) These relationships were already explained in Chapter 3 herein.

2. *Substantive rationality* (R_c) is concerned with questions of 'content', 'substance', or 'knowledge' which guide the outcome of actions over a certain universe of discourse. For example, economic theory is substantive rationality which is used to resolve problems of economic nature. Rules of law, such as statutes and legal precedents, are the legal rationality which is used to resolve issues of legal substance when deciding legal matters. Other substantive rationalities can be visualised such as technical, social, etc.

3. *Procedural rationality* (R_p) is concerned with the choice of procedures by which decisions over the universe of discourse are taken. The choice of actions by the controller (CR) at the object level and by the metacontroller (MCR) at the metalevel depend on both substantive and procedural rationality. As an example, decisions in courts of law can be cited, where rules of law can be identified with the former and legal procedures with the latter. Procedural rationality is a concept proposed, in a slightly different sense, by Simon[9] and discussed by Kickert.[1,12]

4. *Evaluative rationality* (R_e). Decision making is goal-orientated. Without entering into a discussion of the origin or the teleogical orientation of organisations (that is, whether goals 'exist' or are 'attributed' to decision makers), evaluative rationality refers to either *the goals* toward which decision makers appear to strive and/or b) *the criteria* by which goal attainment is defined and evaluated. Kickert[1] suggests a differentiation of evaluative rationality depending on whether complete or incomplete information about the preference ordering of goals prevails.

In summary, structural rationality (R_s) is concerned with *structure*, evaluative rationality (R_e) with *ends*, substantive rationality (R_c) with *content* and procedural rationality (R_p) with *form*.

THE DATA, INFORMATION, INTELLIGENCE TRILOGY

A distinction needs to be made between *data, information, and intelligence.* To do so, we make the following assumptions and consider:

(a) *data*, as the input stimuli in the form of signals and messages entering the cognitive processes. (See Figure 2).

(b) *information* is that portion of input stimuli retained as knowledge and

(c) *intelligence* is that portion of information used by the decision processes for decision making or action.

Figure 3 illustrates the differentiation between the three terms and shows how information is a subset of data, and how in turn intelligence is a subset of information. A similar differentiation was suggested by Murray.[13]

Fig. 2. The differentiation between data, information, and intelligence.

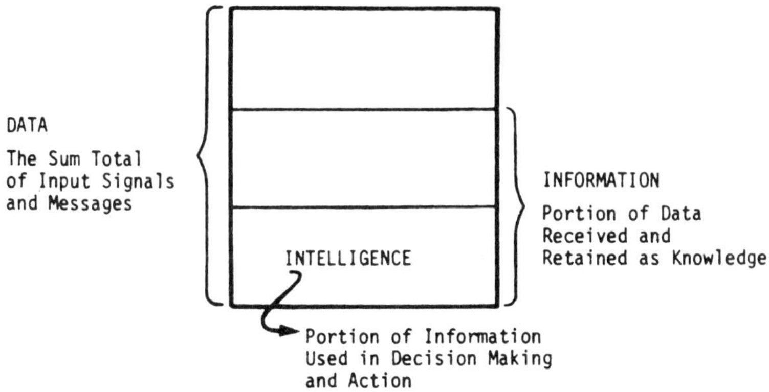

Fig. 3. Information is a subset of data and intelligence is a subset of information.

In our model, *cognitive processes* consist of the mental functions which act to separate information from data, and *decision processes* that convert information into intelligence. Cognitive and decision processes can be aggregated to constitute *behaviour*. (See Figure 4). No pretense is made here to identify particular brain functions with particular cognitive or decision processes.

It is interesting to note that, when *intelligence* is used to formulate a decision, it enters another system as *data*. Indeed it can be said that *the result of one system's decision becomes another system's data*, lending a reflexive property to the data-information-intelligence cycle as follows:

(i) A decision can be used in the same system to formulate another decision.

(ii) Behaviour is reflexive in that a system can act upon its own decision(s)

(iii) Even when the decision resulting from the use of intelligence in one system is used in another system as data, the communication cycle, which embraces data-information-intelligence, can be modelled isomorphically by utilising a simple system with feedback. For all intents and purposes, cognitive and decision processes can be collapsed in one system as shown in Figure 4. This oversimplification overlooks the intricacy of models such as those proposed by Royce and Buss.[14]

Usually the term *data* is fairly well understood because it is used in the sense of

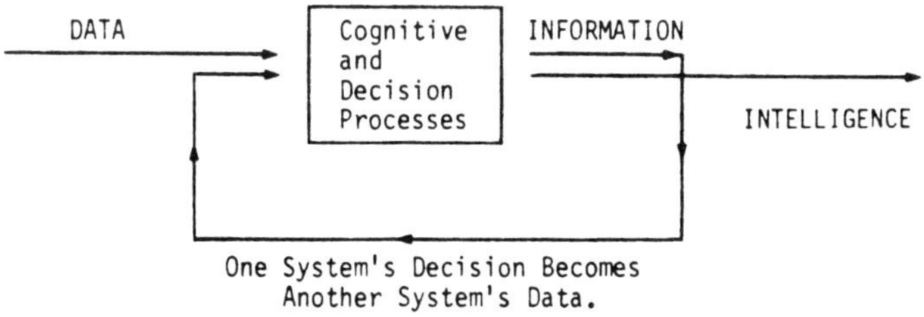

Fig. 4. How data becomes intelligence.

'raw data' which connotes the kind of material to which, as yet, no specific use has been assigned.

Information is the most widely used term. It has been associated with both data and with intelligence. For purposes of exposition, it is hoped that the three terms can be kept distinct from one another, with *information* as more comprehensive than *intelligence* which is reserved to signify that portion of information which is applied to decision making.

RELATING DATA-INFORMATION-INTELLIGENCE TO RATIONALITIES

The differentiation between data, information, and intelligence can be made clearer when relating these concepts to the four types of rationalities defined earlier. The term rationality can describe either data, information, or intelligence, although for the most part:

1. Data can be characterized as substantive rationality (R_c) and procedural rationality (R_p). It is *substantive* because it describes *content*. It is *procedural* because it describes *form*. However, it lacks evaluative rationality (R_e) because it is not concerned with particular ends. It also lacks structural rationality (R_s) because it is not necessarily related to any specific control system or metasystem.

2. Information may contain procedural rationality (R_p), substantive (R_c) and structural rationality (R_s). However, given that it has not been directed to any specific action, it is void of evaluative rationality (R_e). Thus, it lacks ends toward which a posterior decision is directed.

3. Intelligence must contain the four types of rationality (R_s, R_c, R_p and R_e) to deserve this name. The subset of information chosen for decision making and directed to a particular use by the introduction of evaluative rationality (R_e), becomes intelligence.

If we study the elements of the control system – E, CS, CR and MCR, at any level of recursion – only the controller and the metacontroller are able to make decisions which lead to meaningful action. The CS receives its evaluative rationality (R_e) from either the controller (CR) or the metacontroller (MCR) and thus cannot, by itself, produce intelligence. The controller (CR) can take actions after intervention from the MCR which supplies structural rationality (R_s). The metacontroller (MCR) is the agent which supplies structural rationality (R_s) to the controller at the object level. In turn, the metacontroller is subject to the rationalities provided by its own metacontroller at higher levels of recursion. See chapter 19.

RELATED MODELS

While not explicitly mentioning the distinction between data, information and intelligence, many authors have referred to it tangentially. Maruyama[15] concerned himself with the 'metaorganisation of information' or the organisation of how we look at the universe. He postulated the existence of three metaorganisations or 'universes' which lead to corresponding conceptualisations of information:

1. *Classificational universe* with classificational information
2. *Relational universe* with relational information, and
3. *Relevantial universe* with relevantial information.

Maruyama argued that most of the evidence available to us falls in the relational universe where we are mainly concerned with finding mutually exclusive categories with an emphasis on hierarchical arrangement. Evidence, in the classificational universe, is 'knowledge about something'. The relational universe is event-orientated, 'substance is not even assumed'. This universe is concerned with how an event or individual relates to another event or individual. The emphasis lies on how one affects the other and the direction of influences. Finally, in the relevantial universe, information relates to how much it affects each individual, it is affective-oriented, depends on what the interests and the significance of a subject are for a particular individual.

Maruyama[16] further extended his work by introducing the concept of 'mindscapes' or 'epistemological patterns' which shape the cognitive map or style of all individuals. The mindscapes act as metaorganisers of our conceptualisation of the world. They influence our rationalities and hence our decision making. According to Maruyama, while these patterns vary from culture to culture, they pervade the way of thinking of most people in a particular society. See chapter 6, this text.

If we apply Maruyama's[15] earliest characterisation to differentiate between data, information, and communication, we find that:

-*data* reflects a classificational universe, but cannot be used to depict a relational or relevantial one.

-*information* can at best be used to describe a classificational and relational conceptualisation of the world, but

-*intelligence* comprises the elements which make it relevant or relevantial.

Maruyama's[15] three universes can be compared to Locker and Coulter's[17] view of systems which also parallel three basic features of language:

1. *a representative function* by means of which something, e.g. a system can be described.
2. *an expressive function* by means of which an activity of behaviour is manifested, and
3. *an appellative function* by which it is used by one subject to influence another.

Maruyama's classificational universe and classificational information rely mainly on a language's representative function, one that can be used to describe its subject matter. The relational universe requires the use of a language's appellative function because it is concerned with the relationships among individuals and the universe. Finally, the relevantial universe probably needs both the appellative and expressive function because it is concerned with a universe in a way which is only 'relevant' to a particular individual and to his/her special needs. In this case, one is not so much interested in the relationships among individuals, but on how they are related to one's own concerns.

If we apply this view to data, information, and intelligence, we find that:

-*data* can be used for the representative function
-*information* can also be used for the representative function
-*intelligence* is the only one that can perform all three functions of language – the representative function for description, the expressive function to denote activity or behaviour, and the appellative function to show influence or prescription.

Battista's[18,19] seven levels of consciousness, except the first, require some form of evaluative rationality (R_e) and a reflexive function in order to convert information into intelligence. Thus, perception, which is defined as the determination of the meaning of sensation, requires interpretation of the information stemming from it by a higher level of consciousness, that is, that of a metalevel. The meaning of information which originates from each of the first five states of consciousness can only be interpreted by its respective metastates. This view is consistent with the cybernetic view that evaluation requires intervention of the metasystem. Mason and Swanson[20] have used the viewpoint of Semiotics, the theory of signs, to characterise two approaches of measurement: (a) traditional scientific measurement, and (b) management measurement. This characterisation closely resembles the data-information-intelligence distinction presented here:

1. Given that measurements can be considered as signs that carry quantitative evidence, they are akin to *data*.
2. Measurement, as the assignment of numerals and numbers or quantitative signs to reflect properties are akin to *information*, in the sense defined herein.
3. Finally, measurement, as quantitative signs which managers utilize to influence users, is similar to what we defined as *intelligence*. Goal-orientated measurement contains the evaluative function and, as such, it can only be expressed in terms of *intelligence*.

This characterisation of measurement is similar to the differentiation between descriptive and prescriptive statistics and between descriptive and prescriptive theory. Mere description does not carry evaluation and does not indicate the direction of action that will influence the user. For prescription, we need evaluative rationality to obtain goal-directed behaviour. Prescription and action can only be carried out through the use of intelligence. Whereas the observer operates at the descriptive level, the designer formulates recommendations at the prescriptive level.[21–24]

The data-information-intelligence trilogy is useful to characterise management-information systems (MISs) as noted by Murray.[13] The structure and evaluative function of these systems can only be described in terms of intelligence which originates at metalevels. The principle is implicitly used in the methods of structured design which apply to decision support systems (DSSs). The role of the evaluative function in DSSs has been described by Keen and Morton.[25]

SYSTEMS FAILURES AND DYSFUNCTIONS

Systems failures have been the subject of concerted study from several quarters. We recall the work of the Systems Group at Open University[26–30] and the works of Persson.[31] The main purpose of this work has been to approach failures and disasters in a holistic manner to discover emerging patterns that may lead to a prescriptive theory on how to avoid them. The present essay, which outlines the data-information-intelligence distinction plus the indentification of the four rationalities, can be applied to formulate a metasystem approach to diagnose systems failures. As such, it is related to the work on failures described above.

How do systems fail? The full scope of a taxonomy of systems failures is the subject of other papers[32,33] and of subsequent chapters in this text. For the purposes of this presentation it will suffice to outline some of the principles which must hold in order for systems to achieve their goals.

Rationalities hold the key to successful decision making that achieves its goals and hence does not fail. The study of systems failures shows that, in each case, the explanation for dysfunctions lies in the lack of appropriate rationality(ies) or, what is synonymous, of appropriate intelligence. In and by itself, this statement does not carry great new wisdom but points to a way of looking at failures, one which can provide the systems analyst with a direct and insightful approach to diagnose *where and why* systems fail. These principles are the basis for a metasystem paradigm by which failures in systems can be studied.

1. Principle of structure

At each level of recursion, the structural rationality of the control system C is formulated by the metacontroller (MCR). In other words, the structure of a system, defined as the set of relationships between subsystems, aspect systems, and phase systems, is always the responsibility of the metacontroller. The relations between the WHO's, the WHAT's and the WHEN's must clearly be identified, defined, and tested.[1,2] This principle stems directly from the definition of metasystem or from the relationship between object and metalevels in the control system. (See Chapter 3, this book).

This principle can be tested by identifying the various components of the control system under study and recognising their relationships. An important point of contention is whether a recognisable metacontroller exists at each level of recursion. Many times, the metacontrollers are not identifiable or are not explicitly known. This void leads to (a) lack of designed structure, (b) systems without formal or explicit lines of organisation, (c) reduced effectiveness of decision making, and (d) potential failure.

Example: Decision Making in Educational Administration in the Netherlands. Kickert[1] recounts a case study concerning the procedure by which the maximum number of admissions to institutions of higher learning is decided upon in the Netherlands. This study centered upon one particular metadecision, namely structural. The study showed that structural metadecision making was almost nowhere consciously performed. In a vague sense, Parliament in its legislative function, and the Co-ordinative Council of all universities, while developing proposals for the improvement of the process structure, perform this function. However, metadecision making is rather implicit than explicit. This is also a conclusion which can be drawn from the analysis of several other cases,[33] where the functions of the metacontroller have never been formalised.

2. Principle of Legitimacy

To be considered legitimate, evaluative rationality (R_e) must originate in the metasystem. This principle is in keeping with the belief that only a higher level system can arbitrate conflicts among lower level systems, particularly due to the need to formulate these conflicts in a metalanguage which can be understood by all lower level systems and which is neutral with respect to each system in dispute. We recall that evaluative rationality (R_e) refers to the goals and the criteria by which goal attainment is measured.

To qualify as intelligence, this component of the data-information-intelligence trilogy must of necessity contain evaluative rationality. Given that the evaluative rationality (R_e) can only originate in the metasystem, the intervention of the latter is indispensable to the formulation of intelligence.

This principle is not always used in the design of systems. Many systems failures can be traced to the lack of evaluative rationality and of intelligence or to the existence of conflicting signals from the metacontroller concerning the goals which an organisation should follow.

Example: U.S. Aviation Safety System. The National Traffic Safety Board's[34] report on the accident of a PSA aircraft showed that at some point the traffic controller ignored the conflict alert warning. The Report states that the approach controller's decision to ignore the warning was based on 'his judgment and experience'. This failure may be attributable to lack of proper *evaluative rationality (R_e) and intelligence* concerning the direct and opportunity costs of the trade off between notifying or not notifying the parties involved that a warning alert had sounded. The criteria and judgment by which this decision was taken, was obviously in error.

3. Principle of Closure

Intelligence must precede decision making, that is to say, intelligence and the four rationalities which define it, must be present for goal-orientated action to take place. As a corollary of this principle, a decision maker cannot attain his goals unless the communication which is used to implement the decision involves all four rationalities. To say it in another way, decision making based either on *data* or on *information* (as defined herein) is not complete: it must be based on *intelligence*.

Example: The Concorde Fiasco. Building the Concorde was obviously an economic failure.[35,36]. There is no need here to go into the details of the case. Its builders may have struck a technological success but failed to understand some of its spillover effects, not the least of which, was the problem of sonic boom. The Concorde builders and promoters may have had all the *data* and *information* they desired on the technical aspects of the problem, but failed to anticipate the environmental and social consequences of their design. In this sense, they did not hold appropriate *intelligence* to warn them that design changes were necessary, or that the public outcry on the matter would eventually ground the aircraft. They used the wrong evaluative rationality or assigned erroneous weights to decision cues. The metasystem failed, in that it did not formulate the suitable cost-benefit formula by which the sonic boom problem could be evaluated and resolved to the satisfaction of all parties concerned.

4. Principle of Disjunction

Disjunction is a term used by Turner[28,29] to describe a situation when different parties in complex systems do not hold the same evidence. For our purposes, we must define disjunction as the lack of intelligence or of the set of the four rationalities which are comprised in it. For effective metacontrol and to avoid systems dysfunctions and failures, consensus (the converse of disjunction) about intelligence must exist. Disjunction cannot only stem from a lack of consensus, but can also be the result of failures in translation and transduction in the communication channel as explained by Beer.[4]

Disjunction occurs when the decision-making process requires intelligence and only receives data or information. The process by which data and information is transformed into intelligence is not well understood and deserves study at the micro-level.

Example: Open University.[27] The classic example of disjunction was furnished by the Hixon accident which involved a large truck 148 feet long which carried a very heavy transformer and which was hit by a train at a level crossing in England in 1968. The truck moved at 2 miles per hour and in no way could clear the crossing in the 24 seconds warning time before the train appeared. The main ingredient of the tragedy was adjudged to be *information disjunction* which occurred when the individuals involved 'failed to bring together creatively the information they all had or had access to . . .' . . . [they] 'had received the information necessary to avert the disaster, but failed for a number of reasons to consider it actively'. As Turner[28] noted, failure to abide by the *principle of disjunction* is a common known ingredient in many known mishaps.

5. Principle of Compatibility

This principle refers to the set of data, information, and intelligence which managers use as a base for their decisions. Decisions based on world views which do not agree with reality will fail to reach anticipated goals. It is therefore imperative that world views, translated in terms of their respective rationalities, reflect an appropriate understanding of what the real-world system is like. As suggested by Turner[28] there is a lack of compatibility in complex systems when individuals have access to a slightly different set of evidence (data), construct different theories about what is happening (information), and harbour different views about what needs to be done to solve the problem (intelligence). By extension, when a conflicting set of rationalities is used by several parties involved in a situation, it is bound to lead to failure.

Example: National Transportation Safety Board.[34] As the evidence shows, in the case of the PSA aircraft accident, a clear conflict of jurisdiction occurred when the pilot of one plane was, for a while, relying on evidence provided by one traffic controller and then told to exercise visual separation on his own. Visual separation can be construed as control of the flight by the crew itself. Thus the aircraft's operation was controlled by two different controllers, clearly an *incompatibility*. As was also evident from the post-accident report, the substantive rationality (R_c) guiding the two systems was not the same, in particular, knowledge of the relative position of the involved aircrafts with respect to one another. The traffic controller's theory or conceptualisation of what was happening was different from that of the flight crew in the aircraft. Further analysis of the report shows that divergent or conflicting procedural rationality (R_p) was also evident. The *principle of compatibility* was clearly flaunted.

THE METASYSTEM PARADIGM APPLIED TO THE DIAGNOSIS OF SYSTEMS FAILURES

A metasystem approach to information brings out in the open the distinction between data, information, and intelligence. Together with the principles formulated above, it can be used to diagnose complex systems and anticipate potential causes for failures. The metasystem paradigm used in diagnosis involves the following steps:

At each level of recursion:

- Identify each *control system* C, C', C'', etc.
- Identify *components* (MCR, CR, CS, E).
- Differentiate between *data, information,* and *intelligence* flows.
- Identify the origin and destination of the *four kinds of rationality*.
- Identify major foci of *decision making*.
- Diagnose, in each case, whether the appropriate rationalities for making decisions are available and identify voids therein.
- Trace and diagnose cases of systems *failure* to the lack of one of the four needed rationalities.
- Simulate decision-making situations and test availability of appropriate rationality in the control systems under study.

Applications of this paradigm to diagnose systems failures can be found in van Gigch[32] and the chapters that follow. The paradigm was used to study malfunctions of the aviation safety system, the products recall system, the social security system, the child custody legal system, and medical malpractice among others.

REFERENCES

1. W. J. M. Kickert, *Organisation of Decision Making* (North Holland, Amsterdam, 1980).

2. W. J. M. Kickert, and J. P. van Gigch, 'A Metasystem Approach to Organisational Decision Making', *Management Science* **25**, 1217–1231 (1979). See also Chapter 3, infra.
3. S.Beer, *Platform for Change* (Wiley, New York, 1975).
4. S. Beer, *The Heart of Enterprise* (Wiley, New York, 1979).
5. S. Beer, *Brain of the Firm* (Wiley, New York, 2nd ed., 1981).
6. J. P. van Gigch, 'A Methodological Comparison of the Science, Systems, and Metasystem Paradigms', *Int. J. Man-Machine Studies* **11**, 651–663 (1979). See also Chapter 2, infra.
7. Rationality is a concept that appeared, in 1947, in H. A. Simon's doctoral thesis and in all of his subsequent works.
8. H. A. Simon, *Administrative Behaviour* (Free Press, New York, 3rd ed. 1976).
9. H. A. Simon, 'From Substantive to Procedural Rationality', In: *25 Years of Economic Theory*, edited by Kastelein and others (Nijhoff, Leiden, 1976).
10. A. C. J. de Leeuw, 'The Control Paradigm as an Aid for Understanding and Designing Organisations', *Proceedings of the Third European Meeting on Cybernetics and Systems Research* (Vienna, 1976).
11. A. C. J. de Leeuw, 'On Coordination and Autonomy', *Proceedings of the Fifth European Meeting on Cybernetics and Systems Research* (Vienna, 1980).
12. W. J. M. Kickert, 'A Structural Theory of Organisational Decision Making' In: *Current Topics in Cybernetics and Systems*, J. Rose (ed.) (Springer, Berlin, 1978).
13. T. J. Murray, 'Data, Information and Intelligence in a Computer-Based Management Information System', *J. Applied Systems Analysis* **6**, 101–106 (1979).
14. J. R. Royce and A. R. Buss, 'The Role of General Systems and Information Theory in Multi-Factor Individuality Theory', *Canadian Psychological Review* **17**, 1–21 (1976). Reprinted in *General Systems*, **24**, 185–205 (1979).
15. M. Maruyama, 'Metaorganisation of information', *Cybernetica* **4**, (1965). Reprinted in *General Systems* **11**, 55–60 (1966).
16. M. Maruyama, 'Mindscapes and Science Theories', *Current Anthropology* **21**, 589–608 (1980). See also Chapter 6, infra.
17. A. Locker and N. A. Coulter Jr., 'A New Look at the Description and Prescription of Systems', *Behavioural Science* **22**, 197–206 (1977).
18. J. R. Battista, 'The Holographic Model, Holistic Paradigm, Information Theory and Consciousness', *Re-Vision*, Summer/Fall, 99–102 (1978).
19. J. R. R. Battista, 'The Science of Consciousness" In: *The Stream of Consciousness: Psychological Investigations into the Flow of Private Experience*, edited by K. Pope and J. Singer (Plenum, New York, 1978). See also Chapter 10, infra.
20. R. O. Mason and E. B. Swanson, 'Measurement for Management Decision', *California Management Review* **21**, 70–81 (1979). Reprinted In: *Measurement for Management Decision* edited by R. O. Mason and E. B. Swanson (Addison-Wesley, Reading, Massachussetts, 1981).
21. G. Nadler, 'A Systems Engineering Approach to Securing Real World Changes', *J. Applied Systems Analysis* **6**, 89–100 (1979).
22. G. Nadler, 'A Timeline Theory of Planning and Design', *Design Studies* **1**, (July 1980).
23. J. P. van Gigch, *Applied General Systems Theory* (Harper and Row, London and New York, 2nd ed., 1978).
24. J. P. van Gigch, 'Theory Formulation' In: *Research Methodology* (Sacramento, California. Copyright 1979, 1982).
25. P. W. Keen and M. S. Scott Morton, *Decision Support Systems*, (Addison-Wesley, Reading, Pennsylvania, 1978).
26. Open University, *Systems and Failures*, Systems Performance: Human Factors and Systems Failures Unit 1, A Third-Level Course (Milton Keynes, England, 1976).
27. Open University, *The Hixon Analysis Catastrophe and Its Preconditions*, Systems Performance: Human Factors and Systems Failures Unit 2/3 and 4 (Milton Keynes, England, 1976).
28. B. A. Turner, 'The Organisational and Interorganisational Development of Disasters', *Administrative Science Quarterly* **21**, 378–396 (1976).

29. B. A. Turner, *Man-Made Disasters* (Crane Russak, New York, 1979).
30. V. Bignell, P. Peters and C. Pym, *Catastrophic Failures* (Open University, Milton Keynes, England, 1977), and V. Bignell and J. Fortune, *Understanding Systems Failures*, (Manchester University Press, 1984).
31. Bo Persson (ed.), *Surviving Failure, Patterns and Cases of Project Mismanagement* (Almqvist and Wiksell International, Stockholm and Atlantic Highlands, New Jersey, 1979).
32. J. P. van Gigch and W. J. M. Kickert, 'Applications of the Metasystem Paradigm to Organisational Decision Making' In: G. E. Lasker (ed.), *Applied Systems and Cybernetics*, Proceedings of the International Congress on Applied Systems and Cybernetics, Acapulco, Mexico, 1980 (Pergamon, New York and Oxford 1981).
33. J. P. van Gigch, 'Diagnosis and Metamodelling of Systems Failures' In: C. W. Churchman (ed.), *Health of Organisations* (To Appear).
34. National Transportation Safety Board, *Aircraft Accident Report P.S.A.*, B-727, N 533 PS and Gibbs Flite Center Cessna 172, N 77 11 G, San Diego, California, September 25, 1978 (Washington D.C., 1979).
35. B. Lundberg, 'Supersonic Deception', In: Bo Persson (ed.), *Surviving Failures, Patterns and Cases of Project Mismanagement* (Almqvist and Wiksell International, Stockholm and Atlantic Highlands, New Jersey, 1979).
36. A. Wilson, 'The Concorde Fiasco' In: B. Persson (ed.) *Surviving Failures, Op. Cit.* (1979).

19. RATIONALITIES AND METARATIONALITIES IN ORGANISATIONAL DECISION MAKING

J. P. van Gigch

SUMMARY

The metasystem paradigm elaborated in previous papers[1–3] and in Chapters 2, 3, and 18 in this text, is applied to decision making in real world situations.

Decision making is regarded as a multi-level process taking place in an inquiring system which converts inputs (evidence) into outputs (decisions) and whose epistemology consists of rationalities and metarationalities. The model is applied to several examples to identify (a) the problems of concern (universe of discourse), and (b) the rationalities and metarationalities required to solve these problems at each organisational level. This approach can be used to trace and diagnose informational dysfunctions which lead to systems failures.

THE TRADITIONAL ORGANISATION HIERARCHY

Anthony's[4] traditional framework of an organisation with operations, tactical, and strategy/policy levels can be used to study the flow of rationalities for decision making.

We recall that an organisation can be characterised as consisting of three basic levels (from bottom to top).

1. Operations level.
2. Tactical level.
3. Strategy/policy level.

Decision making at each of these levels differs due to differences in:

1. Purpose and function.
2. Problem domain, problem structure, and time horizon.
3. Control and performance measures.

1. Purpose and Function

The functions of the three levels can be characterised by using Beer's[5] Five Systems.

The *operations level* is the same as System One. It accomplishes the main function for which the organisation exists (for example the manufacturing plant in a manufacturing company). It is the only operational unit which has the independent life of a viable system.

The *tactical level* integrates the functions of Beer's Systems Two and Three. System Two is anti-oscillatory, that is, it intervenes to dampen possible oscillations which may arise when operational units operate on their own, without possessing the requisite variety needed to counter the movements of parallel units. System Two does not explicitly exist – it must be designed into the organisation. System Three is the system which provides cohesiveness to the organisation and 'variety-interconnections' to absorb the interaction of elemental units at the operational level. Systems Two and Three at the tactical level regulate the internal environment of the enterprise ('the Inside and Now').

The *strategy/policy level* integrates the functions of Beer's Systems Four and Five. These two systems are designed to manage Ashby's Law of Requisite Variety[6] in the relationships of the organisation with the outside environment ('the Outside and Then').

2. Problem Domain, Problem Structure and Time Horizon

Given that the purpose and function of each organisational level is distinct, the problems, domain, and structure dealt at each level are different.

The operations level is designed to provide direct support to the daily activities of production. In Simon's[7] terms, problems at this level are usually well structured, repetitive, programmable, prespecifiable, and suitable for the application of algorithms.

At the top, the strategy/policy level deals with predominantly novel, ill-structured, non-programmable issues, 'non-specifiable' and suitable for heuristic thinking. Here, problems have usually long-term horizons as most are related to future-orientated planning issues.

The domain, structure, and time horizon of problems at the tactical level fall in between the description given for the operations and strategy/policy levels. Refer to Table 1.

3. Control and Performance Measures

Differences in the type of measures used at the three levels are related to the types of issues described above. At the operations level, criteria used for control are direct and clear-cut, whereas those used to evaluate progress and performance at the strategic/policy levels tend to be indirect and fuzzy. Another way to characterise this difference, is to note that the evaluation at the operations level is more suitable to *quantitative measurement*, whereas the strategy/policy level relies more on *qualitative measurement*. Both quantitative and qualitative forms must be accepted in the context of measurement of social science domains.[8,9]

EPISTEMOLOGY OF THE INQUIRING SYSTEM

The control paradigm provides a model of the decision-making process whereby the controller (CR) is seen as exercising directed influence over the controlled system (CS). It is useful to elaborate the model and describe further the transfer function which, at each level determines the output y. It can be visualised as an *inquiring system*, in the sense given to this term by Churchman[10] and van Gigch.[11] An inquiring system is a process or 'activity which produces knowledge'. In this case, it converts evidence into decisions. This process of inquiry is guided by its *epistemology* which consists of the four types of rationality which guide inquiry. The transfer

Table 1 Differentiating between organizational decision making at the operations, tactical, and strategy/policy levels

Level	Problem Domain	Purpose/Function	Problem Structure	Horizon	Control and Performance Measures
Strategy/Policy	Policy Decisions Planning Future-Oriented Decisions	To support the formulation of the policy.	Not-prespecifiable. Novel. Ill-structured. Non-programmable. Suitable for heuristics.	Long-Term	Mission is formulated in terms of broad performance measures.
Tactical	Management Control	To formulate plans to implement policy.	In between ill-structured and well-structured spectrum.	Medium-Term	Broad performance measures are operationalized.
Operations	Operational Control	Accomplishment of established plans.	Pre-specifiable. Programmable. Suitable for algorithms.	Real-Time. On-Line.	Evaluation in relation to predefined criteria.

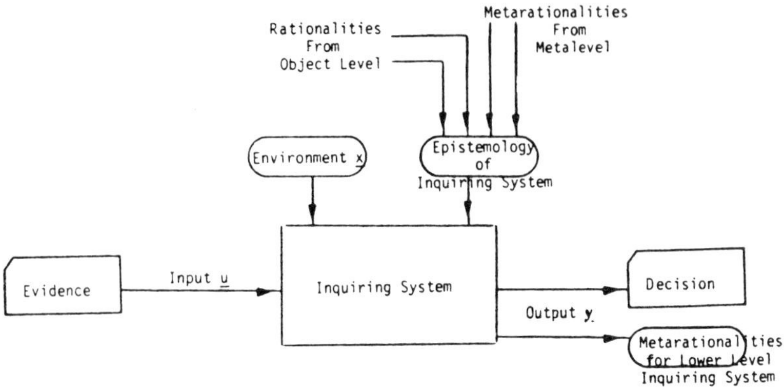

Fig. 1. Organizational Decision Making Viewed as an Inquiring System with Epistemology.

function of the inquiring system can be written as:

$$y = f(x, u, e)$$

with y = output
u = input
x = environment
e = epistemology $f(R_s, R_e, R_c, R_p)$

Figure 1 describes how the black-box concept is elaborated as a decision making process: the inquiring system processes *input* (evidence) into *output* (decisions). The inquiring system's epistemology consists of the four rationalities defined by van Gigch[12] where $e = (R_s, R_e, R_c, R_p)$. Some of these rationalities originate at the metalevel and some at the object level. The inquiring system of upper levels produce *metarationalities* for the epistemology of lower-level inquiring systems as shown in Figure 2, where the traditional organisational levels of the firm – operations, tactical, strategy/policy, and metapolicy levels – are depicted. This multilevel decision-making model shows that, at each level, the epistemology of the inquiring system obtains its rationalities either from the object level or from the metalevel. Usually, substantive (R_c) and procedural (R_p) rationalities originate at the object level, whereas structural (R_s) and evaluative (R_e) rationalities originate at the metalevel. Given their origin, the latter deserve the name of *metarationalities*.

THE DECISION PROBLEM AND CONTROL LEVELS

It is misleading to assume that perfect isomorphism exists among systems levels. As a matter of fact, significant differences from level to level exist among:

1. The universe of discourse.
2. The epistemology which guides the inquiring system (i.e., rationalities and metarationalities).
3. The evidence (in the form of data, information and intelligence).
4. The decisions made at each level.

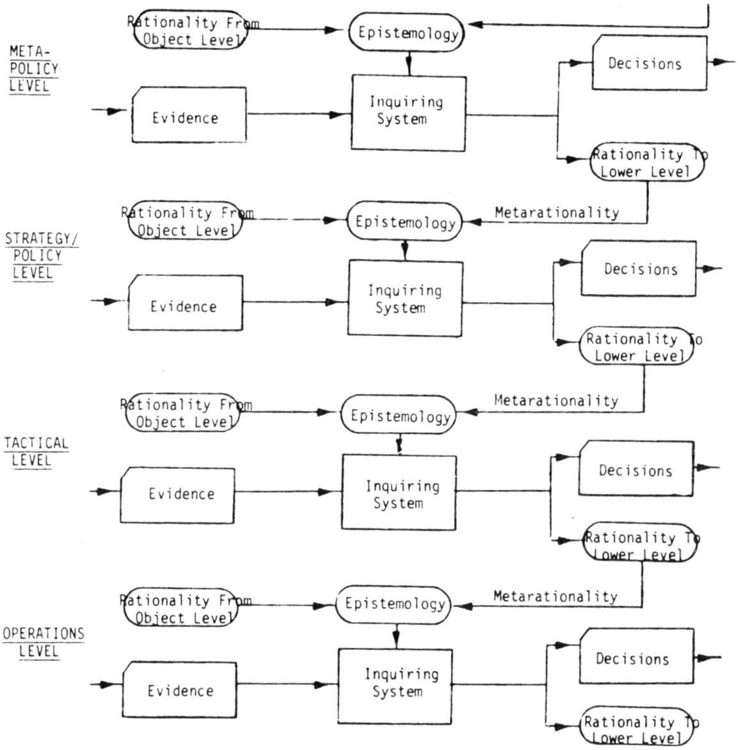

Fig. 2. Generalized Model of Multilevel Decision Making Showing the Flow of Rationalities and Metarationalities.

This article elaborates and illustrates these differences. We proceed to superimpose the control model of decision making upon the organisational framework which consists of the operations, tactical, and strategy/policy levels.

We then obtain the Generalised Model of Multilevel Decision Making as illustrated in Figure 2. It shows an inquiring system at each organisational level. The epistemology of the inquiring system at each level consists of rationalities which originate at the object level and metarationalities which originate at the metalevel.

It is important to note that the *problem* or *universe of discourse* changes at each level of the organisational hierarchy.

In the notation adopted by Mitroff and Betz,[13] the decision theory problem (d.t.p.) takes place at the object level while the metadecision theory problem (m.d.t.p.) takes place at the metalevel. The object level and the metalevel not only involve different orders of logic but, as will be made evident in subsequent illustrations, a different set of evidence and rationalities and, as a result, different decision outputs.

The application of the metasystem paradigm to organisational decision making, consists of analysing decision inputs and outputs, in particular, of identifying the rationalities and metarationalities which make up the evidence (data, information, and intelligence) used to make decisions at each level. See for instance ref. 14.

As a first step, one tries to define the problem of concern or universe of discourse and relate it to the corresponding organisational and control levels as shown in the five examples in Tables II–VI which illustrate, respectively a manufacturing organisation, a budgeting decision, the world of science, the relationship among morality, normative ethics, and metaethics, and finally the hierarchy of epistemological levels.

Table II A Manufacturing Organization

Organizational Levels	Control Levels	Problem of Concern or Universe of Discourse
Strategy/Policy Level	Meta-Meta Level	To decide on product line and products to manufacture.
Tactical Level	Metalevel	To decide product specifications within established product line.
Operational Level	Object Level	To follow specifications when making product(s) and meet schedule.

Table III A Budgeting Decision

Organization Levels	Control Levels	Problem of Concern or Universe of Discourse
Strategy/Policy Level	Meta-Meta Level	To decide upon alternative methods of improving profitability.
Tactical Level	Metalevel	To decide on alternative methods of allocating and budgeting resources.
Operations Level	Object Level	To allocate resources and meet an established budget.

Table IV The World of Science

Organization Levels	Control Levels	Problem of Concern or Universe of Discourse
Philosophy of Science	Meta-Meta Level	Evaluation of scientific paradigms and of epistemologies.
Pure Science Level	Metalevel	New discoveries and the advancement of knowledge.
Applied Science Level	Object Level	The application of knowledge to the production of goods and services.

Table V Morality, Normative Ethics, and Metaethics

Organization Levels	Control Levels	Problem of Concern or Universe of Discourse
Metaethics	Meta-Meta Level	To decide how ethical judgements are defined.
Normative Ethics	Metalevel	To decide upon criteria and norms by which the morality of actions can be evaluated.
Morality	Object Level	To decide whether a particular person's acts satisfy ethical criteria

Table VI Hierarchy of Epistemological Levels

Organizational Levels	Control Levels	Problem of Concern or Universe of Discourse
Meta-Epistemology	Meta-Meta Level	Decisions on how to decide epistemological questions.
Epistemology of the Inquiring System	Metalevel	Decisions concerning the elaboration of methods to generate evidence, to reason, to guarantee truth, and to diagnose dysfunctions.
Inquiring System	Object Level	Operational decisions.

ANALYSIS OF RATIONALITIES AND METARATIONALITIES

In light of the differences in problem domain or universe of discourse which prevail, it is necessary to study the flow of rationalities and metarationalities at each organisational level.

The following table illustrate the inputs and outputs, and the corresponding rationalities in three separate examples. The universe of discourse of these three examples was already described in Tables II, III and IV.

1. Manufacturing Concern

The concepts presented on the previous pages can be used to describe the operations, tactical, and strategy/policy levels of a company which manufactures steel pipe. For this example, the metapolicy level is not applicable. It is clear that the *universe of discourse* at each of these levels is different, for instance at the operations level, decisions are made concerning production scheduling (*when to make* the product), given information from the tactical level where decisions are made about *how to make it*, with information from the strategy/policy level about *what to make*. See Table II

Level	**Universe of Discourse**
Strategy/Policy	*What* to Make
Tactical	*How* to Make
Operations	*When* to Make

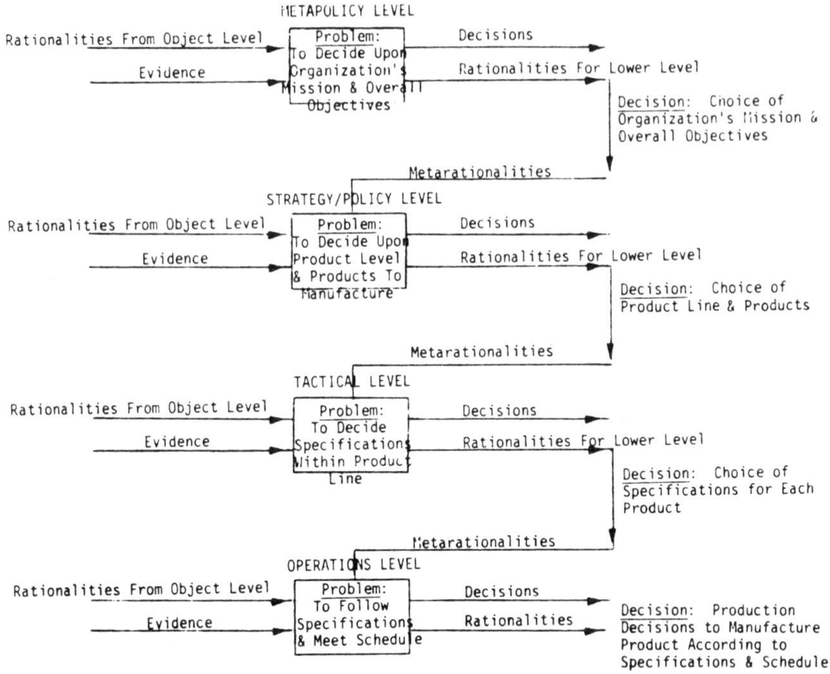

Fig. 3. Application of Metasystem Model to an Example of Production Management.

It is evident that not only the universe of discourse but also the epistemology, or rationalities and metarationalities required to make decisions at each level are different. In other words, the set of rationalities $e = (R_s, R_e, R_c, R_p)$ changes from level to level. Indeed making decisions about *what to make*, is different from making decisions about *how to make*, which, in turn, is different from decisions about *when to make* the particular product in question. Figure 3 and Tables VII, VIII and IX illustrate the origin and destination of rationalities in a manufacturing concern at the operations, tactical, and strategy/policy levels, respectively.

2. A Budgeting System

When a firm or a government agency intends to implement a budgeting system, the *process* by which decisions are taken at each level of the organisation is isomorphic across levels. However the *object of discourse and the epistemologies* which guide the inquiring systems at each level are not. Figure 4 illustrates the three organisation levels (from bottom to top).

1. Operations level.
2. Tactical level.
3. Strategy/policy level.

Table VII Inputs, Outputs, Origin, and Destination of Rationalities at Operations Level of Manufacturing Concern

Universe of Discourse: To follow production specifications and meet a schedule.

Inputs	Description	Origin
R_s Structural Rationality	Instructions on how to organize the structure of the operations level.	Metalevel (Tactical or Strategic Level)
R_e Evaluative Rationality	Criteria by which operational objectives such as cost, output, quality, and time objectives are determined.	Metalevel (Tactical or Strategic Level)
R_c Substantive Rationality	In this case, technical rationality: "How to produce steel pipe," i.e., specification of production methods by which steel pipe can be manufactured.	Metalevel (Tactical Level and Object Level)
R_p Procedural Rationality	Methods and procedures by which specifications are to be implemented.	Metalevel and Object Level

Outputs	Description	Destination
Product Decisions	Decisions concerning the operations and production of steel pipe.	Object Level

Table VIII Origin and Destination of Rationalities at Tactical Level of Manufacturing Enterprise

Universe of Discourse: To decide product specifications within established product line.

Inputs	Description	Origin
R'_s Structural Rationality	Instructions on how to organize the structure of the tactical level.	Metalevel (Strategic Level)
R'_e Evaluative Rationality	Criteria by which the efficient allocation of available resources (raw materials, tools, equipment, and personnel) can be achieved.	Metalevel (Strategic Level and Object Level)
R'_c Substantive Rationality	Managerial and technical knowledge and expertise.	Object Level
R'_p Procedural Rationality	Methods by which managerial and technical expertise are applied to the problem of determining production specifications.	Object Level

Outputs	Description	Destination
Structure (R_s), Evaluative (R_e), and Substantive (R_c) Rationalities for Lower Level	Decisions on structure, evaluation criteria and production specifications.	Operations Level

Table IX Origin and Destination of Rationalities at Strategy/Policy Level of Manufacturing Enterprise

Universe of Discourse: To decide on product line and of products(s) to manufacture.

Inputs	Description	Origin
R_s'' Structural Rationality	Instructions on how to organize the structure of the strategic level.	Metalevel (Board of Directors or Management Consultants)
R_e'' Evaluative Rationality	Criteria by which profitability, return on investment, etc., of enterprise can be evaluated.	Metalevel
R_c'' Substantive Rationality	Economic theory and knowledge of economic climate, markets, and competition.	Metalevel and Object Level
R_p'' Procedural Rationality	Methods by which substantive rationality is applied to make choice of product line and products.	Object Level

Outputs	Description	Destination
Structural Rationality for Lower Levels (R_s', R_s)	Structure of organization.	Tactical and Operations Level
Evaluative Rationality for Lower Levels (R_e', R_e)	Criteria by which lower levels can determine efficient allocation of available resources.	Tactical and Operations Level
Choice of Product Line and/or Products to be Manufactured		Tactical and Operations Level

A fourth level – the meta policy level – can also be included. It may or may not pertain to the same organisation as the other three.

The problems of concern and the outputs of the decision process at each of the three levels are shown in Table III.

Once the problems have been determined, the kinds of rationalities which are required at each level of inquiry can be identified. The structural and evaluative rationalities always stem from the metalevel. The substantive rationality can originate in both, whereas the procedural rationality is usually the product of the object level. The rationalities at each level are illustrated in Figure 4 and Tables X, XI and XII.

3. Philosophy of Science, Pure Science, and Applied Science

The metasystem paradigm can be applied to characterise the process by which pure and applied science produce knowledge.

The various *universes of discourse* and outputs at the various levels of the hierarchy of science were identified in Table IV.

It is obvious that the inquiring systems and the epistemology which guides these inquiring systems differ from level to level.

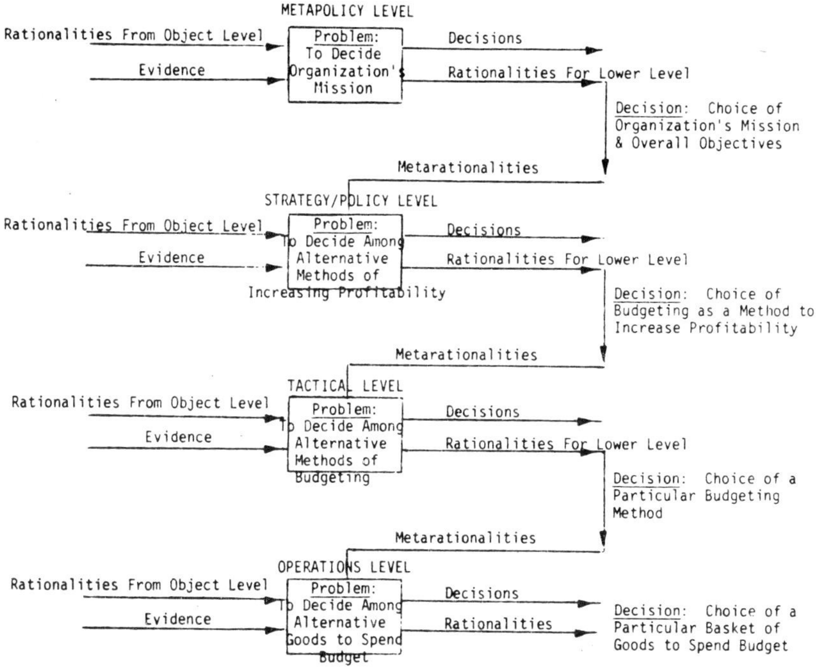

Fig. 4. Application of the Metasystems Model to a Budgeting System.

Table X Budgeting System Inputs, Outputs, Origin, and Destination of Rationalities at Operations Level

Universe of Discourse: To allocate resources and meet an established budget.

Inputs	Description	Origin
R_s Structural Rationality	Instructions on how to organize the structure of the operations levels.	Metalevel (Tactical or Strategic Level)
R_e Evaluative Rationality	Comparative price and utility of available goods and materials used in operations.	Metalevel (Tactical or Object Level)
R_c Substantive Rationality	Knowledge in Accounting Theory, Breakeven Analysis, Price and Utility Theory.	Object Level
R_p Procedural Rationality	Procedures by which cost and value of different commodities and materials which could fit the budget can be compared.	Object Level

Outputs	Description	Destination
Operating Decisions	Decisions concerning the allocation of commodities and raw materials that meet the budget constraints.	Object Level

Table XI Budgeting System Inputs, Outputs, Origin, and Destination of System Rationalities at Tactical Level

Universe of Discourse: To decide on alternative methods of allocating and budgeting resources.

Inputs	Description	Origin
R'_s Structural Rationality	Instructions on how to organize the structure of the tactical level.	Metalevel (Strategic/Policy Level)
R'_e Evaluative Rationality	Criteria by which effectiveness of alternative methods of allocating and budgeting can be assessed.	Metalevel (Strategic/Policy Level)
R'_e Substantive Rationality	Knowledge of Accounting and Financial Management.	Object Level
R'_p Procedural Rationality	Procedures by which effectiveness of alternative allocation and budgeting methods can be compared.	Object Level

Outputs	Description	Destination
Structural (R_s) and Evaluative Rationality (R_e) for Operational Level	Decisions concerning how to establish operating budget.	Operations Level

Table XII Budgeting System Inputs, Outputs, Origin, and Destination of Rationalities at Strategy/Policy Level

Universe of Discourse: To decide upon alternative methods by which profitability can be improved.

Inputs	Description	Origin
R''_s Structural Rationality	Instructions on how to organize the structure of the strategic/policy level.	Metalevel (Board of Directors or Outside Consultants)
R''_e Evaluative Rationality	Criteria by which alternative methods of increasing profitability can be compared (e.g., return on investment).	Metalevel
R''_e Substantive Rationality	Knowledge of Economic Theory, Capital Budgeting, Financial Analysis, etc.	Object Level
R''_p Procedural Rationality	Procedures by which alternative methods of increasing profitability can be compared.	Object Level

Outputs	Description	Destination
Structural (R'_s) (R_s), Evaluative (R'_e) (R_e), and Substantive (R'_e) (R_e) Rationalities for Lower Levels	Decisions on structure, on evaluation criteria, and methods to improve profitability.	Tactical and Operations Level

At the *consumer/user level* the inquiring systems must convert technical knowledge obtained from the output of the applied science level into goods and services. A typical example of this type of system can be found in business and industry whose tasks are orientated to the conversion of practical knowledge into saleable products. At this level, the goal is minimisation of cost or maximisation of profit, quality, output, and so forth. The inquiring system receives part of its rationality, in the form of metarationalities, from the applied science level which is the metalevel for the consumer level:

R_e (evaluative rationality) in the form of criteria by which the goals of cost, profit, and quality of goods and services produced can be evaluated.

R_c (substantive rationality) in the form of knowledge on how to manufacture those items.

R_p (procedural rationality) in the form of methods and procedures by which substantive rationality is applied and used at the consumer level.

At the *applied science level*, the conversion system transforms the outputs of pure science and tries to find applications which can be manufactured at the consumer/user level. The processes are known as the diffusion of inventive activity and technology transfer. At this level, the epistemology of the inquiring system consists of metarationalities and rationalities which are obtained from the metalevel and object level, respectively.

R'_e (evaluative rationality) criteria by which the economic and technical feasibility of manufacturing certain goods and sources are evaluated.

R'_c (substantive rationality) which consists of scientific, economic, and technical knowledge by which the conversion from pure to applied science are effected.

R'_p (procedural rationality) by which the above rationality R'_c can be applied.

At the pure science level, the aim is to push the boundaries of knowledge and to make discoveries. To accomplish this objective, the inquiring system at this level requires metarationalities which originate in the Philosophy of Science level which acts as the metasystem for the pure science level.

R''_e (evaluative rationality) by which progress of knowledge can be evaluated and by which priorities to rank the importance of the various fields of research activity can be established, based on their relative probabilities of success.

The head of a laboratory devoted to pure science is always weighting the expected payoffs of different projects, to choose the ones which hold the best promise of results. Even pure science cannot afford the luxury of undirected inquiry.

R''_c (substantive rationality) consists of all knowledge accumulated in the branch of science or discipline under study. This rationality originates, for the most part, at the object level.

R''_p (procedural rationality) consists of paradigms or methods and procedures by which scientific inquiry is conducted. As an example, the choice of refutation over verification, as a criterion to assess the progress of science, is guided by a sort of procedural rationality.

At the Philosophy of Science level, the subject of inquiry consists of evaluating the comparative advantages of different paradigms of science in order to provide lower levels with reasoning processes and criteria by which the progress of knowledge can be evaluated. The outputs of this level have already been described as inputs to the lower level inquiring system – that of the pure science level. The philosophy of Science inquiring system uses paradigms to evaluate paradigms which it has developed within its own level or which may have originated in other sciences or disciplines.

Structural rationalities (R_s'', R_s', R_s) were not mentioned in any of the inquiring systems described above. In all cases, structural rationality is concerned with providing norms and guidelines by which *the structure* of the organisation is established. It is always provided by the metasystem. The hierarchy of levels in the organisation of science does not necessarily coincide with the administrative hierarchy which manages the organisation. Structural rationality is a managerial concern rather than a scientific concern and would, therefore, be obtained from the metalevel in the administrative hierarchy.

In Chapter 20, the metasystem paradigm is used to analyse the flow of rationalities and metarationalities in the context of the problem of cost-containment in a health-care program. The reader may also feel inclined to analyse other examples such as multilevel ethics in Table V or the hierarchy of epistemological levels in Table VI.

CONCLUSION

Diagnostic and Prescriptive Tool

The procedure outlined in the previous pages leads to the identification of the *problems* (universe of discourse) and *decisions* which are taken at each level of the metasystem. At the same time, it leads to the recognition that, as the problems change, so do the *evidence* and *rationalities* required to solve them. The evidence rationalities and metarationalities required to settle issues at the operations level are different from those needed at the tactical or strategy policy levels, etc. Repeated studies of systems failures reveals that in the past, not enough importance to those differences has been paid.

The said procedure can thus, be used to *describe* ex-post facto what is taking place in the organisation, and may reveal the course of processes which have caused past organisational dysfunctions.[14,15] The procedure can also be used in a *prescriptive* mode to define, at the design stage, the epistemology of the inquiring systems which are needed to define the different universes of discourse of the metasystem. In either case, whether it is used as a diagnostic descriptive of prescriptive tool, the analysis of rationalities and metarationalities provides an effective method to identify breakdowns in information flow and ensure that the appropriate data, information and intelligence needed for effective decision making, are available.

REFERENCES

1. W. J. M. Kickert and J. P. van Gigch, 'A Metasystem Approach to Organisational Decision Making', *Management Science*, **25**, 1217–1231 (1979). See Chapter 3, infra.

2. J. P. van Gigch, 'A Methodological Comparison of the Science, Systems, and Metasystem Paradigms', *Int. J. Man-Machine Studies* **11**, 651–653 (1979). See Chapter 2, infra.
3. J. P. van Gigch, 'The Metasystem, Rationalities, and Information' Chapter 18, infra.
4. R. N. Anthony, *Planning and Control Systems: A Framework for Analysis* (Harvard University Press, Boston, Massachussetts, 1965).
5. S. Beer, *The Heart of Enterprise* (Wiley, New York, 1979).
6. R. Ashby, *Introduction to Cybernetics* (Chapman Hall, London, 1956).
7. H. A. Simon, *The Shape of Automation* (Harper & Row, New York, 1965).
8. J. P. van Gigch, *Applied General Systems Theory* (Harper & Row, New York and London, 2nd edition, 1978).
9. J. P. van Gigch, *Research Methodology* (California State University, Sacramento, 1979, 1982).
10. C. W. Churchman, *Design of Inquiring Systems* (Basic Books, New York, 1971).
11. J. P. van Gigch, *Applied General Systems Theory*, Op. Cit.
12. J. P. van Gigch, 'The Metasystem, Rationalities and Information', *Op. Cit.*, Chapter 18, infra.
13. I. I. Mitroff and F. Betz, 'Dialectical Decision Theory: A Metatheory of Decision Making', *Management Science*, **19**, 11–24 (1972).
14. T. Delacroix, 'A Metasystem Approach to Environmental Pollution Control in Japan', *Human Systems Management*, **4**, 178–188. Reprinted in *General Systems*, **XXVIII**, 17–27, (1983–84).
15. J. P. van Gigch, 'Diagnosis and Metamodeling of Systems Failures', In: C. W. Churchman (ed.), *Health of Organisations* (In process).

20. THE APPLICATION OF THE METASYSTEM PARADIGM TO COST CONTAINMENT IN A HEALTH CARE PROGRAMME*

Ann F. van Gigch

SUMMARY

The metasystem paradigm is used to study the problems afflicting a health care system. The paradigm is applied to identify several malfunctions which stem from the following fragmented solutions due to a noncomprehensive outlook on the part of decision-makers, goal conflicts among different levels of the control hierarchy, inadequate information and rationalities for problem solving, lack of evaluation of results, procedural inflexibility to solve issues of differing complexity and ineffective use of public input and participation. Recommendations for changes to improve the system's organisation and control are made.

INTRODUCTION

The metasystem paradigm can be applied to the California Certificate of Need (CON) Programme, which regulates the construction and expansion of health facilities and health services, to identify dysfunctions or failures in the system. The metasystem paradigm, with its underlying rationalities and principles, has been enunciated by van Gigch.[1] Dysfunctions in the rationalities or disregard for the principles which guide the behaviour and decision-making in a system, lead to dysfunctions or failures.

A metamodel is needed in order to properly define the problem of health care cost containment. In organisations, problems and priorities differ at various levels. The operational level seeks to protect its individual interests, frequently to the detriment of the entire system.[2] Consequently, a metacontroller is needed to define and resolve conflicts among levels of the organisation. In particular, in the health care system, hospitals seek to expand services, while the Office of Statewide Health Planning and Development (OSHPD) seeks to control excessive and duplicated services.

This application of the metasystem paradigm demonstrates its value in the analysis of organisational decision-making. The identification of failure leads to the proposal of recommendations to foster the goals of a CON Programme.

* At the time of printing changes in the CON Programme in California have resulted in the deregulation of hospital equipment purchases and services expansion. Therefore, this chapter should be read as an example of a control system failure without regard to exact representation of the present state of affairs.

BACKGROUND

In response to the increased governmental burden of hospital expenditures post Medicare, the federal government enacted the National Health Planning and Resources Development Act of 1974 (PL 93-641). This law mandated the establishment of the State Certificate of Need (CON) Programme by 1980. It also established health systems agencies (HSAs) and State Health Planning and Development Agencies (SHPDAs). HSAs are federally-designated agencies responsible for health planning for a specified geographical area. They may be non-profit corporations or units of government. A majority of the governing body must be consumers, not involved in providing or financing health care. The intent of Congress was to inject consumer input into local health decision-making. The SHPDA is responsible for state health planning and development. It may have responsibility for the implementation of the CON Programme.

The California Certificate of Need (CON) Act of 1976 was enacted to comply with the requirements of PL 93-641. California first enacted a Certificate of Need (CON) law in 1969. The 1976 legislation set a threshold of $150,000 in capital expenditures which triggered the CON process. Future amendments have raised the threshold to $400,000 for diagnostic equipment and $600,000 for capital expenditures. The amendment of 1981 (SB 930) permits the office of State Health Planning and Development (OSHPD) to waive, at its discretion, public meetings and hearings on CON applications.

California is divided into 14 health service areas, each of which is served by a Health Systems Agency (HSA). The OSHPD is responsible for the development of the California State Health Plan and the implementation of the CON Act of 1976. The agency reports to the Health and Welfare Agency. The Advisory Health Council, a statutory commission, has final responsibility for the State Health Plan and Statewide Health Facilities and Services Plan, a second volume of the broader State Health Plan. This State Health Plan serves as a basis for CON review. The Advisory Health Council, also, has responsibility for the CON appeal process.

Since its inception the CON legislation has been a source of conflict between the hospitals and the regulating agencies. The law has been attacked, unsuccessfully, in the courts as unconstitutional. Antitrusts suits have been brought against HSAs for attempting to induce co-operation between the hospitals in the planning of health services.[3] The relationship between the hospital, on the one side and the HSAs and OSHPD, on the other, continues to be an adversary one.

Although the CON legislation controls capital expenditures of various kinds of hospital facilities, such as acute care and long term care, only general acute care hospitals have been considered in the present study. It is in this category of hospital that the greatest capital expansion occurs with the resultant impact on hospital expenditures.

APPLICATION OF THE METASYSTEM PARADIGM

1. Domain

According to Kickert and van Gigch,[4] the control system of the CON Programme can be visualised as follows. At the first level of recursion in the Certificate of Need (CON)

Programme, the hospital administration is the controller (CR), the hospital services are the controlled system (CS) and the environment (E) includes other hospitals, medical community, the public and medical technology suppliers. The metacontroller (MCR) is the Office of Statewide Health Planning and Development (OSHPD). The control system (C) includes the hospital administration and the hospital services. See Figure 1.

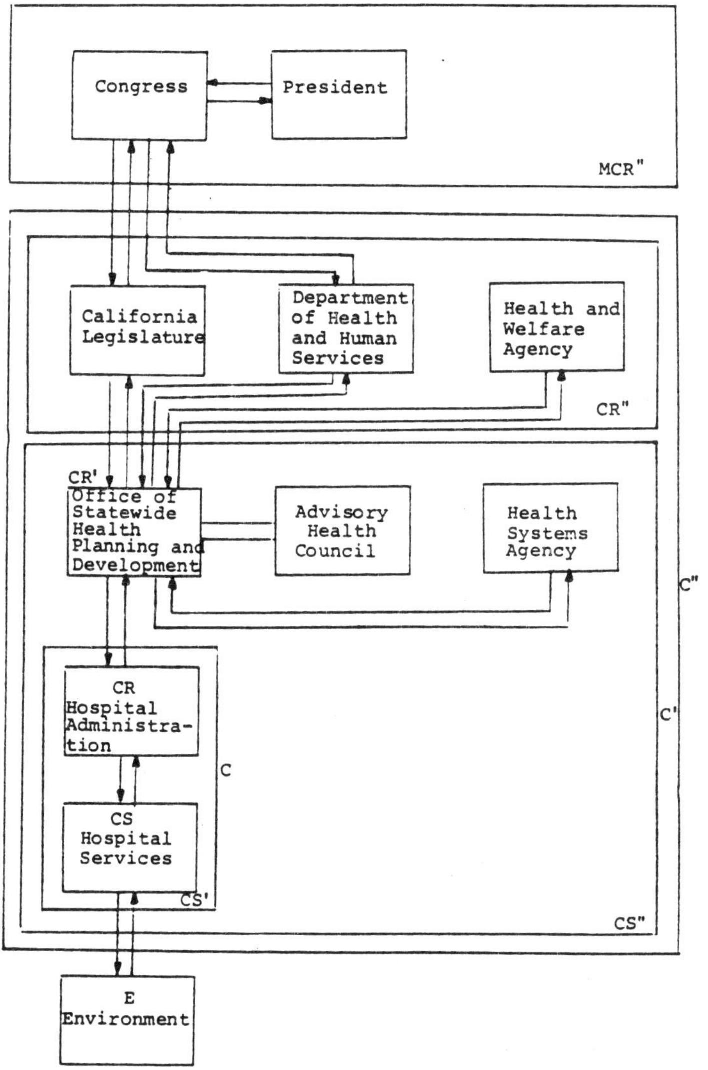

Figure 1.
Three Levels of Recursion in the California Certificate of Need Control System.

The hospital administration and its board of directors are involved in capital expansion and replacement decisions. These decisions aim at meeting the objective to provide quality health care to patients. Another important objective of the hospital is the attraction and retention of the medical staff. The hospital administration must choose between the demands of the department chiefs, who seek to improve their own departments. The hospital administration becomes an arbitrator in a sometimes fierce competition for available resources.

Groups within the environment, impact on the hospital's decision-making. The community demands accessible, quality health care. Other hospitals in the areas compete to provide health services to the same population. Suppliers of medical technology promote the latest advancements in the field.

The OSHPD has designated responsibility for the implementation of the CON Act of 1976. It seeks to control excessive or duplicated health services, as defined by the Statewide Health Facilities and Services Plan. The OSHPD has named the Health Systems Agencies (HSAs) to independently review and recommend approval or denial of CON applications for their designated area of control. The HSAs have both a planning and a regulating function. Every three years, they are responsible for the development of an area health plan. The Advisory Council is responsible for the State Health Plan and CON appeals.

The relationship between the hospital, on the one side and the HSA and OSHPD on the other, is often viewed as adversarial. The conflict can be attributed to the different world views of the regulated and the regulator. With major funding for the HSAs and OSHPD dependent on the goodwill of the federal government, a power imbalance frequently exists between the adversaries in terms of money, manpower and time.

At the second level of recursion, the OSHPD functions as the controller (CR'), the hospital administrator along with the hospital services is the controlled system (CS'), the state legislature acts as the metacontroller (MCR') of a control system (C'), which includes the OSHPD, the hospital administration and the hospital services. (Refer to Figure 1).

The Legislature is concerned with the containment of health care costs. Its role is the enactment of appropriate legislation to control health care costs. These decisions are impacted by various interest groups dedicated to growth, quality health care and regulation. Through the CON Act of 1976, as amended, the procedures, by which the OSHPD regulates expansion of the state's hospital facilities and services, are set.

At the third level of recursion, the Legislature is the controller (CR''), the OSHPD, the hospital administration and the hospital services are the controlled system (CS'') and the Congress acts as the metacontroller (MCR'') of a control system (C'') which includes the Legislature, the OSHPD, the hospital administration and the hospital services. (See Figure 1).

Congress enacts legislation related to hospital facility and services expansion, with which the states must comply. The thrust of this legislation depends on the national health policy. In developing the national health policy, Congress is influenced by the various interest groups.

2. Organisational Levels

The CON Programme can be interpreted as a hierarchy of systems levels. (See Figure 2). At each hierarchical level (operations, tactical, strategy/policy and metapolicy) the

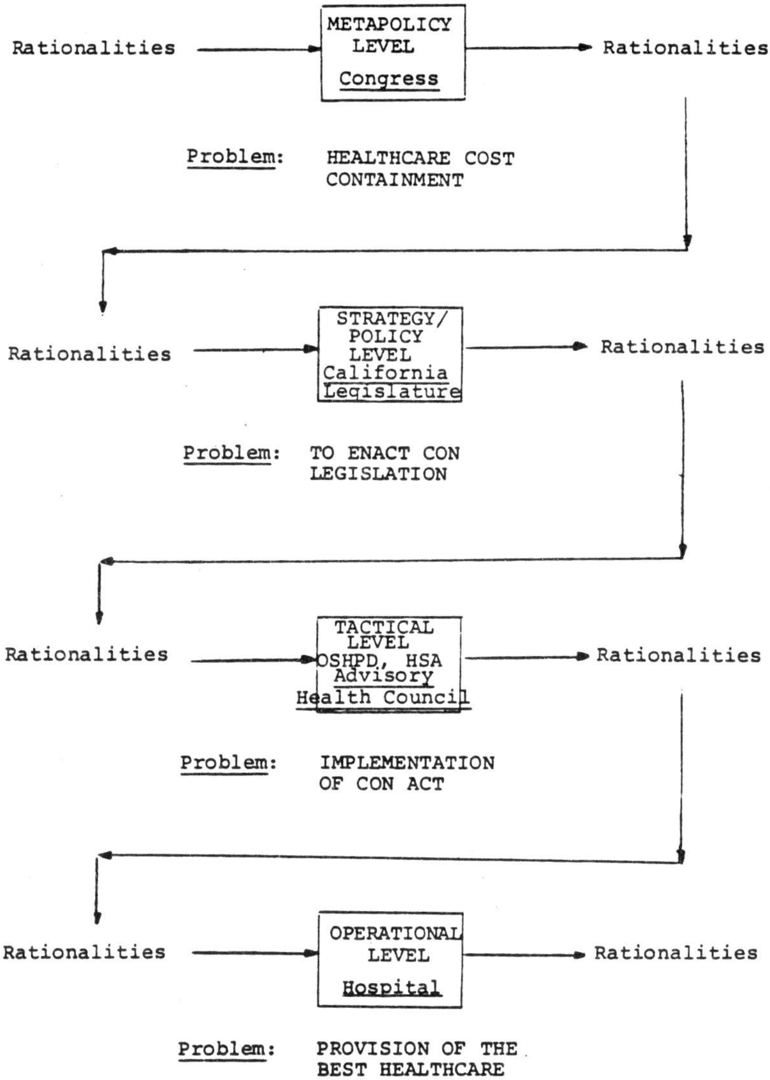

Figure 2.
Interpretation of the CON Program as a Hierarchy of Systems Levels.

rationalities (structural, substantive, procedural and evaluative) are used to make decisions to solve the specific problem at that level. Elaboration on these rationalities follows below. The rationalities that are the outputs from each level are sent to the next level.

At the operational level, the main problem (or universe of discourse) for the hospital is to provide quality care in accordance with the demands of its patients and medical staff in a competitive environment. The hospital has objectives that often conflict with the regulatory plan laid down by the Legislature. These conflicts must be overcome by the OSHPD. Beer's model,[5] for effective organisation of a system, based on an analogy of the control of the brain, contains five control systems. The hospital corresponds to Beer's System One, the viable system, where the regulatory plan is implemented.

At the tactical level, the task of the OSHPD is to direct the implementation of the CON Act, in accordance with the requirements set down by this legislation. The OSHPD correponds to Beer's System Three, in that it administers the CON Programme and provides cohesion to the regulation of facilities and services. The main task of the HSAs is the development of area health plans for their service areas and the review of CON applications. The HSAs appear to correspond to Beer's System Two, in that they act as an anti-oscillatory system for the competing hospitals through the health needs plan for the area.

At the strategy/policy level, the main problem for the Legislature is to develop a plan to regulate the hospital facilities and services to comply with national health policy and legislation (PL 93-641). The Legislature corresponds to Beer's System Five in that it develops the CON Program for the state.

At the metapolicy level, the main problem for Congress is to decide on a method of containing health costs, in line with a national health policy. Policy is developed and a plan designed based on feedback received from various lobbyists, including the

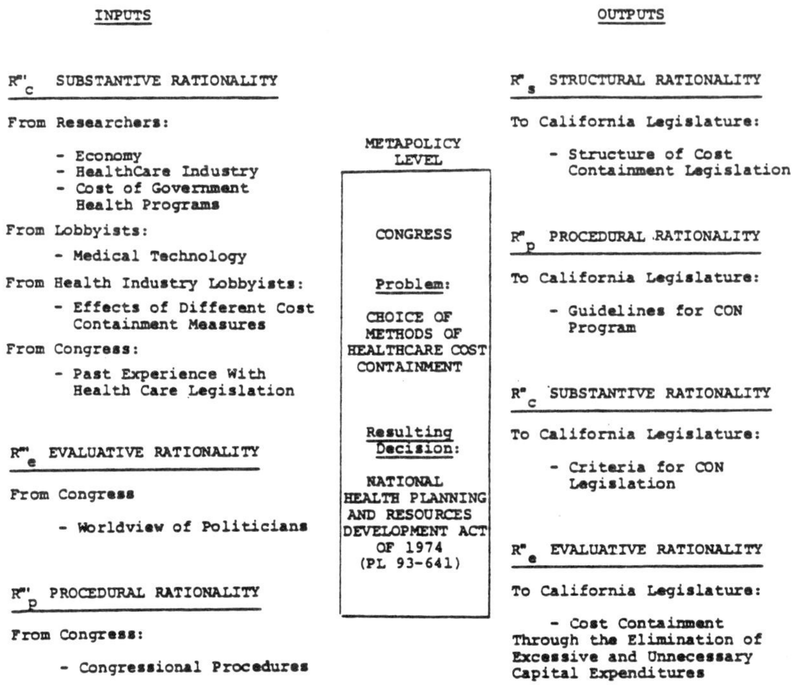

Figure 3.
Rationalities in the CON Program at the Metapolicy Level.

American Hospital Association, the American Public Health Association and the American Health Planning Association and performance audits of the State CON Programme by the Department of Health and Human Services (DHHS). The DHHS and the lobbyists appear to correspond somewhat to Beer's System Four. Congress corresponds to Beer's System Five.

3. Rationalities

At each hierarchical level in the CON system, the origin and destination of the rationalities, which are used to solve the main problem can be identified. Figures 3, 4, 5 and 6 represent a detailed elaboration of the rationalities. In each figure, the reader can identify the components of each of the rationalities. As an example the California Legislature uses substantive rationality (R_c), structural rationality (R_s), procedural rationality (R_p), and evaluative rationality (R_e) to comply with the mandate of PL 93-641 to enact CON legislation. Further, R_c in that case, includes PL 93-641 criteria for CON legislation, the state of the economy, the rising cost of Medi-Cal, the health care industry and past experience with health care containment legislation. The inputs and outputs at each of the four levels are shown in the above named figures.

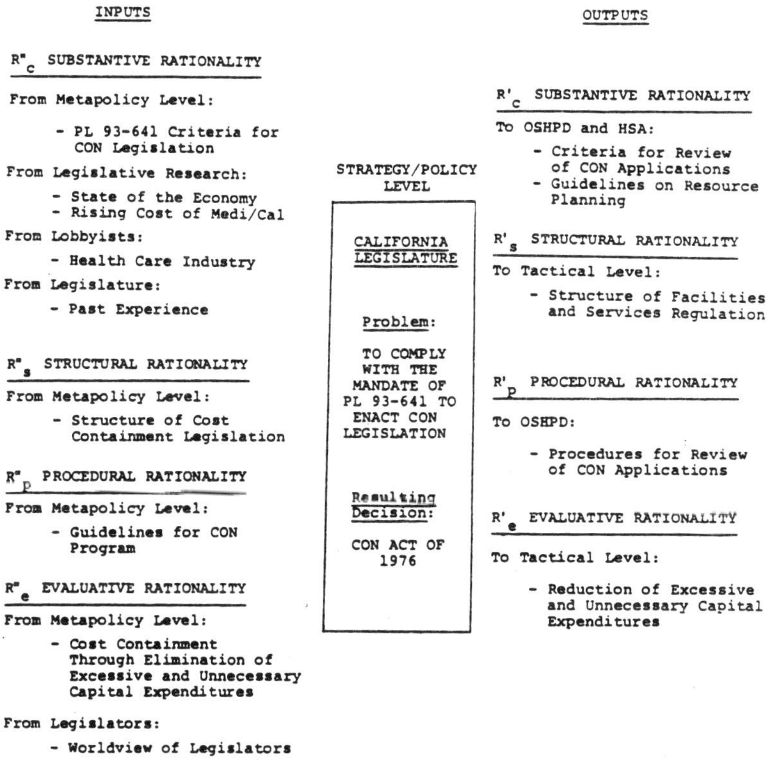

Figure 4.
Rationalities in the CON Program at the Strategy/Policy Level.

INPUTS

R'_c SUBSTANTIVE RATIONALITY

From California Legislature:
- Criteria for Review of CON Applications
- Guidelines on Resource Planning

From Advisory Health Council:
- State Health Plan

From OSHPD:
- Knowledge of Hospital Industry

From HSA:
- Recommendations of HSA on CON Applications

From Hospitals:
- Financial, Managerial, Legal and Planning Theory
- CON Applications

R'_s STRUCTURAL RATIONALITY

From California Legislature:
- Structure of the Facilities and Services Regulation

R'_p PROCEDURAL RATIONALITY

From California Legislature:
- Procedures for Review of CON Applications

From OSHPD:
- OSHPD Methods for Review of CON Applications

R'_e EVALUATIVE RATIONALITY

From California Legislature:
- Reduction of Excess and Unnecessary Capital Expenditures

From Tactical Level:
- Worldview of CON Reviewers and Health Needs Planners

TACTICAL LEVEL

OSHPD, HSA, ADVISORY HEALTH COUNCIL

Problem:

IMPLEMENTATION OF THE CON ACT OF 1976

Resulting Decision:

CON APPLICATION DECISIONS AND CON APPEAL DECISIONS

OUTPUTS

R_c SUBSTANTIVE RATIONALITY

To Hospitals:
- Application Decisions

To OSHPD:
- Appeal Decisions

R_e EVALUATIVE RATIONALITY

To Hospitals:
- Distribution of Health Facilities and Services in the Community Without Excess or Duplication

Figure 5.
Rationalities in the CON Program at the Tactical Level.

4. Requisite Variety

According to Ashby's[6] Law of Requisite Variety, the OSHPD must be capable of generating a variety equivalent to the variety of the hospital. Variety generators are the means by which a system produces variety. Variety amplifiers and attenuators are those means by which a system expands or reduces its variety respectively.[7]

Within the CON system variety generators, amplifiers and attenuators can be identified. The development of a future expansion project by a hospital may entail several years of planning. Advanced cost finding methods and data automation are variety generators for a hospital. The OSHPD generates variety through the State Health Plan and HSAs collaboration on CON applications. Whether the State Health Plan acts as a variety amplifier or a variety reducer depends on the information in the State Health Plan. The OSHPD can cope with the variety of the systems it controls, to the extent that the information it receives from these systems is accurate and comprehensive. If the information in the California Statewide Health Facilities and

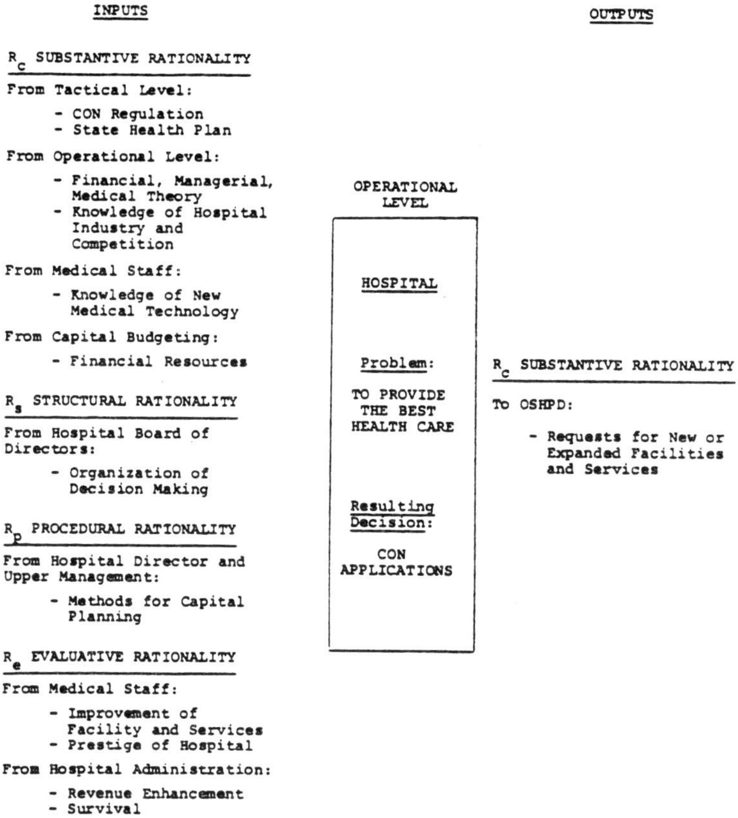

Figure 6.
Rationalities in the CON Program at the Operational Level.

Services Plan is lacking or out-of-date, the OSHPD will require more information (or variety) in order to control the hospital's capital expenditures. The same is true in the case of the CON application. If the latter is not accurate and comprehensive, the variety of the CON application is increased, and to control it, the OSHPD must increase its variety in order to handle the process. Hospitals frequently underestimate the cost impact and overestimate the need of new facilities and services in CON applications, in order to present its demands in the best light. Public meetings, where community input is voiced, are also variety amplifiers to the OSHPD. Knowledge of the effects of similar projects, in other service areas, increases the variety of OSHPD.

Funding is another mechanism by which the variety of the OSHPD can be reduced or amplified. The performance of the HSA and the OSHPD depend to a great extent on federal funding. In 1981, the HSAs funding was cut thirty percent. Certainly, this action reduces the ability of the HSAs to develop quality area health plans and quality review of CON applications.

The behaviour of the OSHPD, in the regulation of facilities and services expansion, is constrained by the CON Act of 1976. The criteria and guidelines for the review of CON applications are set by the CON statute. The variety to control the CON process is thereby limited and the OSHPD needs less variety to control the CON process. On the other hand, if the legislative constraints were less, the OSHPD could develop better criteria to review CON applications. It is conceivable that more power would enable the OSHPD to better control unnecessary and excessive health facilities and services.

SYSTEMS MALFUNCTIONS AND FAILURES

Systems malfunctions and failures can be related to the rationalities and principles of the metasystem paradigm which are explained in chapters 18 and 19.

1. Fragmentation of Solution

The CON programme represents a fragmented solution of the problem of rising hospital costs. Excessive and duplicated hospital facilities and services are only one factor which influences the hospital costs. Other vital factors are the retrospective reimbursement system, third-party payors, physician control of demand and supply, increased intensity of services, medical technological advances and general inflation. These latter factors have not been addressed either by Congress or the California Legislature. The politicians deal with the problem of health cost containment from a perception of reality different from that of other experts. The regulation of facilities and services as a solution, reflects the philosophy of the politicians and the relative strength of the hospital industry versus the public. The hospital industry succeeded in the elimination of the provision for state administered rate controls from PL93-641, prior to passage.

The fragmented solution represents a failure in substantive rationality (R_s) by the Congress and California Legislature. Evaluative rationality (R_e) failed since CON regulation alone could not brake the escalating hospital costs and goals were not met. The *principle of closure* was violated when all the variables in hospital costs were not considered in the solution. There was a lack of compatibility of the evidence used by the politicians in their decision-making.

2. Goal Conflicts

Goal conflicts between the regulators and the regulated is another critical system failure. Control in the CON Programme is made difficult by conflicts of goals between the Congress, State and the OSHPD, on the one side, and the hospitals, on the other. The regulator's goal is to meet the health needs of the citizens, without excessive or duplicated facilities and services. The goals of the regulated (hospitals) are to provide the best health care to their patients, to enhance their prestige as an institution and increase revenues. The goal of the regulators implies a co-operative distribution of health services within a given service area. However, the goals of the hospital are shaped by self-interest. Within a service area, several hospitals may be competing for the same patients. Each hospital wishes to provide the best health care, translated into the latest technological advances. High technology services, such as computerised axial tomographical scanning and cardiac catheterisation labs represent high cost services. Hospitals seek to recover their costs, plus a fixed percentage of gain. Given that the gain in absolute terms will be larger the greater the costs, it is to the advantage of hospitals to provide high cost services. The retrospective reimbursement system is a disincentive to cost containment.

Conflicting goals between the OSHPD and the hospital is a failure of structural rationality (R_s). The structure of the decision-making at the object level (hospital) must be decided by the metacontroller (Legislature). The structure should eliminate the conflict of goals between the hospital and the OSHPD. The failure of structural rationality (R_s) also involves the evaluative rationality (R_c), in that the goals of the hospital and the OSHPD conflict. The evaluative rationality (R_e) provided by the Legislature lacks incentives for the hospitals to control costs. The *principle of legitimacy* was violated when the physicians were not a party to the decision to regulate hospital facilities and services and do not accept the idea of solutions imposed by regulators.

3. Lack of Evaluation

The lack of effectiveness evaluation in the CON Programme represents a grave weakness of the CON Programme. The variables that effect health costs are complex and interrelated. As yet, no method has been devised to measure the relative effect of controlling a single variable, such as capital expenditures.[8]

Performance auditing, by the DHHS, of OSHPD does not assess the degree of cost containment due to CON. Rather, it audits the use of the funds provided by the federal government to the OSHPD.

There is no internal evaluation by the OSHPD of CON decisions, in relation to the elimination of excessive and duplicated facilities and services. To date, there has been no effort to compare CON decisions with the facilities and services needs projections of the State Health Plan. Denied, unnecessary capital expenditures represent savings, through the reduction of assets and potential increases in operating expenses.[9] However, measuring savings from denials, omits savings due to deterred CON applications which result from the presence of the regulatory process.[10] Intuitively, the OSHPD and HSA are aware of the deterrence factor, but it is impossible to measure. According to some regulators, the large percentage of approved applications (approximately 90%) reflects a congruence with the State Health Plan.

The lack of measurement of goal attainment in the CON Programme is a failure of structural (R_s) substantive (R_c) and procedural (R_p) rationalities. The OSHPD and the Legislature may not know how to measure the effectiveness of the CON Programme. Additionally, the evaluation is not a part of the structure or procedures built into the programme.

4. Inadequate Information

A fourth failure in the CON Programme is the inadequacy of information for decision-making. Decision-making in the CON Programme is tied to the needs assessment of the State Plan. Unfortunately, the information necessary for specific CON applications is frequently lacking or out-of-date. The quality of the State Health Plan is not a reflection on the skills or knowledge of the health planners. Although the methodology for sound health planning is readily available, the quality reflects the amount of funds allocated by the federal government, to carry out the mandate for health planning. Total capital investment, by institutional health care providers, was estimated at an excess of $2 billion annual in 1976. To regulate this investment and carry out comprehensive health planning. Congress allocated $30 million in FY 1976, roughly 1.5% of the value of the investment to be reviewed.[11]

Criteria used to evaluate CON applications is vague and ambiguous. Terms such as 'reasonable', 'optimal utilization', and 'adverse effect' abound. The definition of these terms, frequently, depends on the subjective judgement of the decider. The OSHPD fails to require, in verifiable form, crucial financial and operating data. Therefore, CON application information, supplied by hospitals, frequently lacks the quality needed for efficient analysis. The use of statistical averages, to demonstrate the effect of capital expansion on rate increases, underestimates the extent of the increase. To justify the need for a new expansion or service, hospitals are motivated to present information in the most favourable light.

Information disjunctions in the State Health Plan lead to failures of translation and transduction. Vague and ambiguous criteria may result in a failure of substantive rationality (R_c) on the part of the OSHPD, as reviewers interpret the meaning of fuzzy terms. Evaluative rationality (R_e) may also be affected by the lack of clear standards. Ambiguous information from hospitals leads to failures of transduction by reviewers. The variety of the OSHPD and HSAs does not match the variety of the hospitals. To date the OSHPD does not have a computerised data base. Inadequate funding results in an inability of the regulators to effectively review the large number of CON applications entering the system, from throughout the State of California.

5. Inflexibility

Another dysfunction is inflexibility in the CON Programme. All CON applications, until January 1, 1982, be they simple and non controversial or highly controversial, followed the same process through the CON Programme. Resources spent on routine applications, reduces resources for review of controversial applications. The amendment of the CON of 1976 will allow the OSHPD the discretion to omit the HSA public meeting on routine applications. This will speed up the processing of a proportion of applications.

The prior inflexibility of the CON process represents a failure in procedural rationality (R_p) which may now be corrected by the amendment.

6. Ineffective Public Participation

The myth of public participation has been perpetuated to satisfy the original intent of Congress in PL 93-641. The experience of consumer participation, in planning and regulating health facilities and services, has been similar to the experience in other public policy decision-making. The consumers on HSA boards and committees have tended to look to the provider representatives for guidance in decision-making.[12] The intent of the Congress in passing PL 93-641 was to inject consumer participation into the health planning and regulation process. In practice, public participation becomes interest group participation. The erosion of commitment to seek public participation is evidenced in the amendment to the CON Act of 1976 which gives the state the right to waive public participation in 'non-controversial' CON applications.

The regulators have failed to provide the consumer with the skills (substantive rationality, R_c) to participate meaningfully in the process. The regulator's behaviour also represents a failure in procedural rationality (R_p) in that the procedures for educating the consumer are lacking or ineffective.

7. Limited Scope

The scope of the CON Programme is limited to quantity and distribution of health facilities and services. The CON Programme fails to address the issue of the impact of new medical technology. Consequently, the CON Programme has been more effective in reducing bed expansion than new services expansion.[13,14] The question of the overall effect of new medical technological advances on the health care of all the participants in the health care delivery system is not addressed, either by the hospitals or by the regulators. The philosophy of the health care delivery system is based on high technology curative medicine. The result is the prioritisation of financial resources for this type of medical care, to the detriment of preventive and primary medical care.

The personal values of the regulators, related to health care, often conflict with their stated objective of cost containment. This objective can never be reached with a CON Programme alone. Dysfunctions of structure result from a mismatch between the structure of the CON Programme and the government's ability to pay the rising cost of governmental health programmes. As the government seeks to shift the burden to the private health insurers by limiting government reimbursement of Medicare and Medi-Cal, the system's fluctuations increase in its intensity. The cost of one day in the hospital is the variable which will send the CON Programme over the threshold of stability. At some point, other measures to control hospital costs will be demanded by interest groups. The CON Programme may be replaced by another regulatory programme. Alternatively, the CON Programme could be reorganised to incorporate another regulatory measure.

DIAGNOSIS OF SYSTEMS FAILURES

The weaknesses of the CON Programme itself, which include goal conflicts, deficiencies in both information and review criteria, a lack of effectiveness evaluation, inflexibility of procedures, and ineffective public participation, pale beside the failure of the legislative process to meet the challenge of the complex problem of rapidly rising health care cost. A cost containment programme, designed solely around the regulation of health facilities and services expansion, was inevitably doomed to failure. The problem of rising health costs demands a solution that considers all the controllable factors. The development of a public policy on health is hampered by goal conflicts among the participants in the system. There may be no solution to the containment of health care costs within the present fee-for-service, high technology, curative medical care delivery system. Fiscal constraints, at the federal and state level, are forcing the politicians to take some action. The direction of this action will depend on the strengths of the interest groups involved.[15]

A change in the health care delivery system will require a change in values by health care providers, politicians and the public. The change in values will not be easily effected. The question to be explored is the extent to which the proliferation of high technological advances is causing a degradation of the health care system. The answer could be the trigger to accomplish the value change necessary to reform the health care delivery system.

The diagnosis of failures in the CON Programme leads to recommendations for changes in the system which hopefully might lead to goal attainment.

1. Prospective Reimbursement and HMOs.

A more comprehensive approach to health care cost containment is imperative. Legislation is needed to affect additional controllable factors in health care costs. This might include a change in the reimbursement system from the present one which provides hospitals with incentives to increase costs, to one with incentives to reduce costs. The development of Health Maintenance Organisations (HMOs) should be encouraged to decrease the utilisation and intensity of services.

2. Goal Conformity

To bring the goals of the hospitals into conformity with the goals of regulators, the cost containment system should provide incentive mechanisms to hospitals to reduce costs. The incentives might take the form of rate regulation or prospective capitation reimbursement. The former induces the hospital to contain costs, to maximize the difference between costs and the set rate. The latter, integrated with the HMO concept, induces providers to reduce costs by paying them a set amount for health care per person served, regardless of services provided.

3. Accountability

Accountability demands that the CON Programme evaluate its effectiveness. To this end, a method of evaluation of the congruence of CON decisions with the State Health Plan needs assessment should be developed. Also, the effect of CON on hospital assets should be studied. Goal achievement by the OSHPD depends on a sound information base for decision-making. It is imperative that sufficient funds be allocated to develop sound State Health Plans.

4. Adequate Information

The sophistication of the CON review process must be increased so that its variety matches the variety of the hospital systems. The OSHPD must develop objective criteria, free of ambiguity, for the evaluation of CON applications.

5. Public Participation Education

The OSHPD should fulfil the intent of the Congress and the Legislature in regard to public participation in the planning and regulating process. To this end, an educational program should be developed to equip consumers to participate meaningfully in the decision-making process of the CON Programme.

POST-SCRIPTUM

Several recommendations of this study have since been implemented. The change from retrospective to prospective reimbursement in the Medicare Program and fixed-rate-per-day payment systems may result in a slowing of the rate of increase in hospital costs. The new payment systems provide an incentive to hospitals to control unnecessary capital expenditures and to reduce the intensity of services. However, market forces inherent in the prospective reimbursement systems have neither decreased capital expenditures nor reduced bed supply[16]

REFERENCES

1. J. P. van Gigch, 'The Metasystem, Rationalities and Information' Chapter 18, infra.
2. S. Beer, *Brain of the Firm* (Wiley, New York, 2nd. Ed. 1981).
3. L. H. Glanz, 'The Legal Aspects of Health Facilities Regulation', In: H. H. Hyman. (Ed.) *Health Regulation* (Aspen Systems Corporation, Germantown, Maryland 1977).
4. W. J. M. Kickert and J. P. van Gigch, 'A Metasystem Approch To Organisational Decision-Making', *Management Science* **25**, 1217-1231 (1979), Chapter 3, infra.
5. S. Beer, *Op. Cit.*
6. *W. R. Ashby, An Introduction To Cybernetics* (Wiley, Science Editions, New York 1963).
7. S. Beer, *Heart of Enterprise* (Wiley, New York 1979).
8. W. J. Bicknell, and D. Chapman Walsh, 'Certificate-of-Need: The Massachusetts Experience', *New England J. Medic,* **292,** 1054-1075 (1975).
9. U.S. Department of Health and Human Services, Office of the Assistant Secretary for Health, *Development of An Evaluation Methodology For Use In Assessing Data Available To The Certificate-of-Need- (CON) and Health Planning Programmes,* prepared by Arthur D. Little, Cambridge. MA. 02140, (April 1982).
10. D. S. Salkever, and T. W. Bice, 'The Impact of Certificate-of-Need Control on Hospital Investment', *Health and Society* **54,** Arthur D. Little, Cambridge, MA. 02140 (April 1982).
11. E. J. Hanley, 'Regulation of Health Facilities and Services Under Public Law 93-641', In: H. H. Hyman, (Ed.) *Health Regulation, Op. Cit.* (1977).
12. H. H. Hyman, 'Introduction to Key Issues in Regulation of Health Facilities and Services', In: H. H. Hyman, (ed.) *Health Regulation, Op. Cit.* (1977).
13. D. S. Salkever, *Op. Cit.*
14. *J. R. Howell, Regulating Hospital Capital Investment: The Experience in Massachusetts* (National Centre for Health Services Research, Research Summary Series, 1981).
15. L. C. Gawthrop, *Administrative Politics and Social Change* (St. Martin's Press, New York 1971).
16. L. Johns, 'Selective Contracting in California', *Health Affairs,* **4**(3), 32-48, (1985).

21. THE METASYSTEM APPROACH AND THE DYSFUNCTIONAL FAMILY SYSTEM

Nora Femenia

The metasystem paradigm is applied to the study of the family as a control system which operates at several levels of recursion. The role of metacontrollers such as parents, social and governmental agencies, the legal system and the environments is described. Current failures of the family system due to metasystemic dysfunctions are surveyed, in the hope that they may be corrected and overcome.

INTRODUCTION

In their study about middle-class American families, Parsons and Bales[1] drew the following conclusions: there are two basic and irreducible functions of the family. First, the primary socialisation of the children, which permits them to really become members of the society in which they are born and, secondly, the stabilisation of the personality of the adult population, in general.

When we compare these goals with the extent to which these functions are achieved, we realise that there is a high degree of disorganisation of the family. In a great percentage of cases, the family is unable to integrate its members in society. The battered child syndrome, the high percentage of school dropouts, juvenile delinquency and the increasing percentage of heroin addicts reflect serious failures in the functioning of the family. It is urgent to implement a research methodology which permits the identification of these failures and which provides signals of their impending occurrence before they occur. The advantages of prevention are indisputable, because once failures take place, rehabilitation is difficult, costly and, very often impossible. For the purposes of the present study the metasystem paradigm developed by Kickert and van Gigch[2] is used. It provides the metamodel needed to observe phenomena which, while taking place in the family context, must be studied from the point of view of a larger framework.

DEFINITION OF THE FAMILY

The family will be defined as: 'Parents and children, whether or not they live together'.

It is a structure of two elements, the parents, the children and a mutual relationship which derives from the objective of the children's upbringing. We omit from this discussion the large number of special cases in which the level 'parents' suffers all kinds of modifications in the number or in the identity of the adults that fulfil this role. Even in the case of 'surrogate families' (Speck and Attneave[3]) the structure is basically the same.

If we apply the model of control developed by Kickert and van Gigch[2] to the family system, the parents become the controller (CR), the children the controlled system (CS). While, in the model, the controlled system interacts with an environment (E) it is more accurate to describe the environment E as surrounding and directly influencing all levels of the metasystem. Indeed, society constitutes the metaenvironment (bio-social environment) in which the family functions. Therefore, E's influence reaches to all levels, to CR, MCR, MCR' as well as CS.

According to the metasystem paradigm, the control system C at the first level of recursion is made up of CR, CS, E and MCR, where the latter is a metacontroller which controls the controller (CR). Further additions of metacontrollers such as MCR′, MCR″ and so on, define subsequent levels of recursion. With the latter, the level of generalisation and control-domain change: at each level, the controlled systems are different and, at least in theory, additional levels of recursion can be postulated.

While the family system adequately fulfils its objectives, external controls, such as those provided by grandparents, teachers, the family doctor or friends are only needed occasionally. These external controls mediate in functional problems and sometimes fulfil pertinent informational needs. The parental level (CR) is in charge of the family control system and is exercised over the filial level (CS), for the purpose of raising psychological and socially healthy individuals. However, this task is influenced by the action of the environment E over CR, CS and their relation, the actions and messages of CR and the internal (mental) requirements of the functioning of CS which changes, as it develops and matures.

In the ideal system, these variables maintain themselves in a state of controlled fluctuation which do not require the intervention of the metacontroller MCR. Ideally, the objectives are met while the parents can activate the routine control mechanisms: (1) they present responses whose variety is equivalent to the variety exhibited by CS; (2) they modify the internal information processing of CS (considered as a black box) by providing meta-comments which express the extent to which the mental processes of the children depart from the designed target; (3) modify the latter and adapt the reality of CS's goals (in particular in cases of psychological and physical deficiencies to protect handicapped or disabled individuals); and (4) modify CS's environment in order to produce behavioural changes in CS with rewards or penalties.

DOMAINS AND LEVELS OF RECURSION

Given the particular structure of the family, the domain over which the control system C exercises its influence is wide. (Refer to Figure 1). This control system is in charge of all functions of the family: the parents conceive, maintain and rear, orientate the children while they are relatively young. This task is made more difficult due to the biological immaturity of CS. What is in question is the development of shaping of a human mind. Only the external aspects can be made objective and can be measured: weight and height of children, number of hours of care, level of health of the body and indicators of good nutrition. However, there are innumerable items of decisive importance which are concerned with the physical development and cannot be measured. Therefore, one can only describe the influence of CR over CS approximately. The influence that CR exercises while

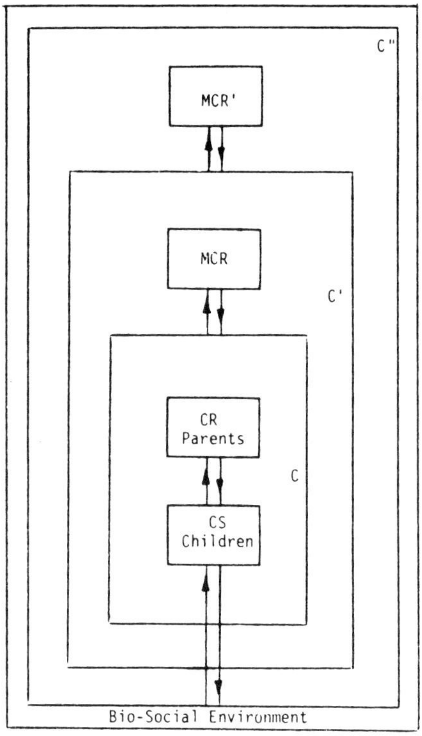

Fig. 1. First and Second Level of Recursion for Family System.

controlling and regulating CS (a human life), can only be described approximately. Furthermore, one has to assume that, from a legal point of view, both parents are responsible and custodian of that life and will try to maintain it. In case they fail (because of abortion, infanticide, abuse, battery or abandonment), the metacontroller will have to intervene. However, intervention occurs only in cases when there exists visible harm, and given the defenseless state of CS, ex-post-facto intervention is not sufficient.

The environment E influences the family system in two ways. Society sends messages on how it believes children should be cared for, protected and taught etc. This operates in the form of a set of socio-cultural prescriptions which guide these tasks. They arrive at CR from two directions: (a) the experience of the parents (consisting of their own childhood recollection of stored learning what their own parents did or did not do to them) and (b) the direct social prescriptions which reach the parents through popular dissemination of psychology, pedagogy, education and so on.

The first level of recursion is made up of the control system C (made up of the subsystem 'parents', the subsystem 'children' and their relationship) and the metacontroller MCR. (See Figure 1). There are several possible metacontrollers, depending on the specific problem exhibited by the family system. For the system with schizophrenic transaction[4], the psychiatric hospital which has custody of mental patients acts as the metacontroller. In the case of abandoned or battered children or precocious delinquents, the juvenile court plays the role of metacontroller. In addition, social agencies such as societies for the protection of battered children intervene. The metacontroller can also be school authorities who deal with the problems of school maladaptation.

The second level of recursion is made up of the previous system (C and MCR) considered as one system C′ with its own metacontroller MCR′. (See Figure 1). The tasks of the metacontrollers must be supported by a body of social laws which are dictated for special cases and then are generalised to others. As an example, we have laws on child custody, family law etc. The body of laws concerned with the family system remedies or attempts to remedy each case on the base of well known models. Occasionally, a new law must be written to cover novel situations. Unfortunately, this method of creating laws leads to a fragmented vision of the family situation which precludes a comprehensive view of the whole family which is needed or desirable.

More often than not, we have a catalogue of prescriptions and check list of possible repairs to solve a litany of failures. There is no prescription for the healthy system. In order to have a clear notion of what constitutes systems failures one must have a clear model of the healthy system. However, the latter is nowhere to be found.

The level of MCR′ (See Figure 2), is fulfilled by the Supreme Court and the Legislature. Here, at the metapolicy level, several questions arise. The first question is related to whether the second level of recursion for the family system under study really exists. Some authors such as Andolfi and Zwerling[5] believe that, in the U.S., no policy at the national level really exists and that serious barriers toward a coherent family policy hinder its implementation. According to these authors[5] a systemic approach to such a policy (which would have the well-being of the individual as its social objective), would require a complete restructing of our society, a job which we are not capable of undertaking because of dearth of skills. The present approach is linear and consists of the adoption of simplistic solutions for individual problems which are strictly isolated from one another. As an example, the problem of drugs is handled by a program specifically designed for drugs and so on. It is very difficult to envisage or implement a systemic approach. For Keniston and the Carnegie Council on children,[6] such a policy could only be carried out in the context of a radical revision of the social structure.

An empty metalevel represents a failure. Another failure resides in the lack of training of the metacontroller to carry out its missions. van Gigch[7,8] has raised the question of whether the theories of one level of the metasystem are isomorphic with that of another. In certain particular cases, such as that of narcissism, they appear isomorphic. Gear *et al.*[9] define the psychic system in conflict as a system with narcissistic interrelationships. Briefly, it involves a kind of double bind where a couple of simultaneous messages mutually incongruent are sent. The whole system functions this way. It cannot be attributed to individual systems. According to Steier, Stanton and Todd,[10] the application of this hypothesis, at the highest level

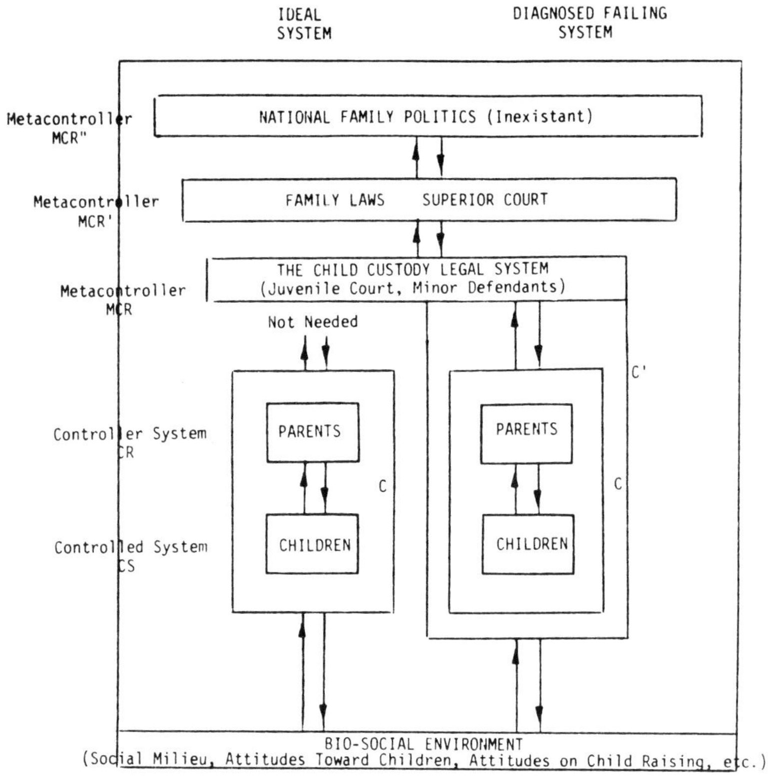

Fig. 2. Comparison of Ideal and Diagnosed Failing System for Several Levels of Recursion.

of the systems C and C′, includes the respective metacontrollers. This is called the 'social transmission' of narcissism.

According to François,[11] it could be argued that the metasystem autogenerates itself starting from subsystemic elements, as well as from the environmental metasystem. The latter will also display a certain failure of rationality in spite of its dynamically stable state. Whether this state is more or less isomorphic with the elements of the operational level is still debatable. However, what is imperative, is to identify MCR and MCR′ (or identify their absence). One could postulate the existence of non-narcissistic MCR and MCR′. However, how can this happen? According to François, given the inherent autogeneration of the human metasystems, the transformations of the family systems, by the introduction in their midst of the needed rationalities, justify the isomorphism between the operational and the metasystem levels and explain the transformation of one toward the other. But how does it occur and by what means? One of the factors in this transformation is a greater knowledge of family dynamics, psychotherapy and thorough change in the family relationships.

SYSTEM FAILURE IN THE FAMILY

According to van Gigch,[12–14] one must first identify failures of regulation and of control.

1. Failures of Regulation and Control

(a) *Viability*. Given that the subsystem 'children' is the main component of the family system, it must be determined whether it receives metasystemic support to ensure its viability. Killing a son, a daughter, abortion, demonstrate failures. As a corollary, does the subsystem 'parents' act to assure the subsystem 'children' of its continued support?
(b) *Antioscillation*. Do any resources exist to compensate for the lack of co-ordination between CR and CS? This antioscillatory function is sometimes performed by grandparents, teachers, social workers, or neighbours who mediate among the subsystems in conflict. It is an essential function.
(c) *Cohesiveness and Variety Matching*. Do the parents have the resources and the knowledge to adapt to the variations which the growth of the children demand from them, or do these variations overcome them? Do they exhibit the appropriate responses and answers to these emerging problems? Do the parents have to remain quiet because they do not know how to answer, or can they resort for support, to resources available in the environment?
(d) *Viable Regulation*. What model of the family do the parents follow? Can it be compared to the family which they have created? Can they see the difference between model and reality and make adjustments? Can they make revisions to the anachronistic model received from their own parents and which is perpetuated beyond justification?

2. Failures in Technology Utilization

While the family presents the image of a mythical support system, a failure can occur when 'home technology' is used to the detriment of outside help or another human contact. According to Harlow's and Zimmermann[15,16] human beings should be brought up by placing a greater emphasis of physical contact over the use of surrogates. The use of television as an electronic pacifier is an example of this displacement.

3. Failures in the Informational and Decisional Processes

(a) *Structural Rationality*. We recall that structural rationality, as defined by Kickert and van Gigch[2] and van Gigch[7] is used to define systems structure (subsystems, aspects systems and phase systems), for example to define *who* does *what* and *when*. One of the most common failures of the family system occurs when parents who, because of their own defects cannot adequately fulfill their role as parents, use the children to carry out responsibilities beyond their age. Therein lie harmful contradictions for the children who must, at the same time, carry out adult responsibilities, while still behaving (according to parents) 'irresponsibly' as children. For a schematic showing the flow of rationalities refer to Figure 3. This behaviour is manifested by assigning to children work that should be carried out by the adults, by allowing them to leave school prematurely or by postponing the independence when they are already in condition to assume it. The metacontroller could control this failure of rationality by not permitting that children be sent to work to support the parents.

(b) *Substantive Rationality*. Failures of substantive rationality in the family context are related to the lack of basic knowledge concerning the dynamics of family relationships and of principles of behaviour. In spite of inordinate efforts to popularise knowledge of psychological principles, it is not exaggerated to assert that what is termed the 'generation gap' can simply be attributed to parents' lack of the basic knowledge of what makes the children 'tick', for example to understand their behaviour, as normal responses to the situations in which they are placed at home, at school and elsewhere. 'The generation gap' is nothing more, nothing less than a lack of substantive rationality or failure to understand 'what is going on'. This is difficult to remedy indeed, because children do not want to listen to parents anyway. They act out of spite and, as a matter of principle, reject the parents advice. Thus the intervention of the metasystem to provide substantive rationality appears imperative in this case.

Errors of substantive rationality can also occur when parents address themselves to the wrong problem or obtain erroneous information about it. This failure has been referred to as the Error of the Third Kind[17].

(c) *Procedural Rationality*. Procedural rationality failures are related to the procedures by which information about actions is transmitted. Failures reside in that these instructions may be confusing or contradictory with the facts (behaviour) of those who provide the instructions, in particular, if these are emitted simultaneously. ('Do what I say and not what I do'). This is the basic notion of 'double bind' referred to by Bateson[18] and invoked earlier as a basic narcissistic mechanism at the individual, social and family levels.

Other failures of procedural rationality may result from the selection of erroneous or counterproductive procedures which are incongruous with the stated objectives, too slow or too costly. The justification for corporal punishement as a way to obtain obedience, as well as authoritative models in the family, have already been discussed for a long time. The generation of children brought up with the permissive laissez-faire suggested by Spock[19,20] has led, due to its results, to a re-evaluation of this model, and to question whether the imposition of certain limits and constraints on children while resorting to techniques of non-corporal punishment may have its value.

(d) *Evaluative Rationality*. The evaluation of the objectives pursued during the children's upbringing should be carried out by CR. This is a little understood subject, given the inability of CR to impartially evaluate its own actions. CR always will tend to rationalise its errors. One would assume that evaluative rationality of this kind should stem from MCR. However, even MCR is partial: no metacontroller exists who has a total and healthy unbiased image of the family from which

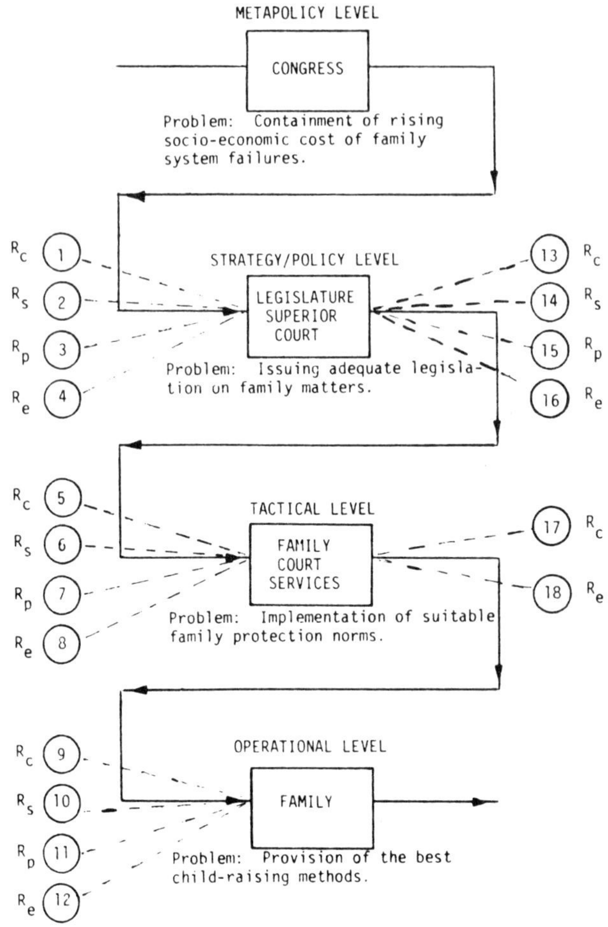

NOTE: For an explanation of the meaning of the rationalities, please refer to legend hereby.

Fig. 3. Levels of Recursion and their Rationalities.

1. National statistics about divorce, battering, etc.
   Social cost evaluation. Scientific Information on family dynamics.
2. Desired family model.
3. Family laws procedures.
4. Legislator's mindscapes.
   Costs of family functions replacement.
5. Case evaluation statistics.
6. Divorce Laws. Child custody laws.
   Regulation on battered children.
7. Auxiliary consultant organizations.
   Prevention of cruelty to children.
8. Reduction of number of children who are state-raised.
   Testing the real effectiveness of tactical level to stabilize the family.

the actions of CR can be judged. Thus in the case when the evaluative rationality does not originate with MCR, a failure in the decision process will take place. Society sometimes provides a substitute for this void by providing other systems from which this rationality can be obtained such as family therapists, family counsellors and others. These supports help the family to leave the narcissistic cycle and to recuperate its true objectives.

(e) *Disjunction*. Occurs when each of the parties has a different set of rationalities: in the process of becoming a family, both parents seem obligated to permanently compare and compete when comparing their male/female, father/mother roles. Feminist movements and other similar ideas have sharpened the need to compare models and reconsider traditional stereotypes, in order to function more harmoniously. However, it becomes very difficult at the level of MCR to take into account new models of the family (non-married parents, family networks, communes) in particular, given the lack of legal precedents for these new forms of family relationship.[21,22]

(f) *Compatibility Dysfunctions* take place when parents have the wrong set of evidence and rationalities about themselves, their children, the family relation or the environment.

4. Failures of Behaviour

We already referred above to incompatibilities between the prescribed rationalities and the observed behaviour. The output message can be congruent or incongruent with the situation; the decider can, on the one hand, state to be in favour of a particular action and, on the other, to do the exact opposite. The action is incongruent with the enunciated objectives and the final influence is confusing and has confusing effects on the recipient.

CONCLUSION

The subject of the evolving death of the traditional family, as it was defined at the beginning of this chapter, is a well known subject. Toffler[23] has imagined new

---

9. Financial resources.
   Knowledge about child-raising.
10. Family planning: when and how to have children.
11. Decision making in the domestic environment.
    Man-wife roles.
    Women's liberation ideas about domestic roles.
12. College evaluation of children.
    Medical and Psychological testing.
    Family dynamics.
13. Revised laws.
14. Special laws: compulsory vaccination, scholarity etc.
15. Codes to guide judge's adjudication.
16. Special laws e.g. about I.Q. requirements or level of instruction to enter schools.
17. Prescription and rules of action for school teachers, juvenile police, religious leaders etc.
18. Degree of mental health of children who are state-raised.
    Evaluation.

models of the family, more adaptable to the social conditions of the future. However, in spite of the well known and current failures of the present family system, it is not yet possible to know how it will be changed and how it will be converted as a result of the present currents in social evolution.

REFERENCES

1. T. Parsons and F. Bales, *Family Socialisation and Intervention Process* (Free Press, Glencoe, Illinois, 1955).
2. W. J. M. Kickert and J. P. van Gigch, 'A Metasystem Approach to Organisational Decision Making' Chapter 3, infra.
3. R. V. Speck and C. L. Attneave, *Family Networks* (Pantheon Books, New York, 1973).
4. M. Selvini Palazolli et al., *Paradosso e Contraparadosso, Un Nuovo Modello Nella Tarapia Della Famiglia* (Feltrinelli Editori, Milano, Italia, 1975).
5. M. Andolfi and S. I. Zwerling (Eds.), *Dimensions of Family Therapy* (The Guildford Press, New York, 1980).
6. K. Keniston, *All of Our Children: The American Family Under Pressure*, The Carnegie Council on Children (Harcourt Brace Jovanovich, New York, 1977).
7. J. P. van Gigch, 'The Metasystem, Rationalities and Information', Chapter 18, infra.
8. J. P. van Gigch, 'Rationalities and Metarationalities in Organisational Decision Making' Chapter 19, infra.
9. M. C. Gear, M. A. Hill and E. C. Liendo, *Working Through Narcissism* (Aronson, New York, 1981).
10. F. Steier, M. D. Stanton, and T. C. Todd, 'Normal and Dysfunctional Families: A Study of Quantitative Distinctions in Organisation', *Proceedings of the 26th Annual Meeting of the Society of General Systems Research* (Washington, D.C., 1982).
11. C. François, 'Consequences of Gödel's Theorem for the Metasystem Paradigm' Chapter 5, infra.
12. J. P. van Gigch, 'Toward Descriptive and Prescriptive Theories of Systems Failures', *Proceedings of the 27th Annual Meeting of the Society for General Systems Research* (Detroit, May 1983).
13. J. P. van Gigch, 'A Taxonomy of Systems Failures and Organisational Dysfunctions', In: C. W. Churchman (Ed.), *Health of Organisations* (In process to appear).
14. J. P. van Gigch and W. J. M. Kickert, 'Applications of the Metasystem Paradigm to Organisational Decision Making', In: G. E. Lasker (Ed.), *Applied Systems and Cybernetics*, Proceedings of the International Congress on Applied Systems and Cybernetics, Acapulco, Mexico, 1980 (Pergamon, New York and London, 1980).
15. H. Harlow and M. K. Harlow, 'Social Deprivation in Monkeys', *Scientific American*, **207**, 137–146 (1962).
16. H. F. Harlow and R. R. Zimmermann, 'The Development of Affectional Responses in Infant Monkeys', *Proc. Amer. Phil. Soc.*, **102**, 501–509 (1958).
17. I. I. Mitroff and T. M. Featheringham, 'On Systemic Problem Solving and The Error of the Third Kind', *Behavioural Science* **19**, 383–393, (1974).
18. G. Bateson, *Steps Toward an Ecology of Mind* (Chandler Publishing, 1972).
19. B. M. Spock, *The Common Sense Book of Baby and Child Care* (Duel Sloan and Pearce, New York, 1946).
20. B. M. Spock, *Dr. Spock Talks with Mothers* (Houghton Mifflin, Boston, 1960).
21. S. E. Raglin, 'The Community Property System in California' California State University, Sacramento, School of Business and Public Administration, Department of Management, Project Research Methodology (Autumn 1982).
22. B. Towne, 'The New California Child Custody Legal System', California State University, Sacramento, School of Business and Public Administration, Department of Management, Project Research Methodology (Autumn 1980).
23. A. Toffler, *Future Shock* (Random House, New York, 1970).

SECTION V

CONCLUDING REMARKS

EXTENSIONS AND FURTHER RESEARCH IN METAMODELS AND METASYSTEMS

Rammohan K. Ragade

SUMMARY

The author offers suggestions for further research on metamodelling and on the metasystem framework and the glimmer of hope that this effort may lead to the emergence of a metaparadigm – 'the grand paradigm' from which all other systems paradigms may one day evolve.

An overview of this book has already been provided by van Gigch in the Introduction. While many of the chapters can be considered a presentation of different aspects of the metasystem paradigm, in order that we may adequately consider extensions and further research, a couple of questions need be asked. The first is whether the metasystem paradigm is justifiable in view of the variety of perspectives requiring the emergence of a new systemic level that can hold together component purposive or teleogenic systems, resolving, co-ordinating and managing their conflicting individual goals. The second is whether it is more appropriate to call the metasystem paradigm a *metaparadigm:* the grand paradigm of all systems paradigms by which all other systems paradigms are to be designed.

Such a question of absolute existence of an entity is bound to failure. The terms are relative. There are no absolute atomic systems as shown by modern subatomic physics.[1] Nor is there a concrete cognition of an ultimate universe of systems, except that posited through a cognitive deductive map. There are many important issues concerning what constitutes information, intelligence and logical systems, or what is freedom of decision-making by actors who are, otherwise, determined by their individual mindscapes.

If the consistent reasoning process within a given logical system is unable to attest to the truth of logically inconsistent statements within the system, then there is a need of an external level system that resolves that inconsistency. Such is the search of the new metasystem paradigm. Its resolution seeks the establishment of alternative perspectives for understanding with different logics, although these may be themselves inconsistent. Wisdom is that knowledge is gained by the acceptance of two or more such perspectives. Voorhees[2] makes a plea for the extension of the existing scientific methodology. He seeks to include, explicitly, the importance of scientific paradigms, due to interplay of the epistemic paths of rationalism, metaphorism, and empiricism as an 'eternal braid'.[3] Human action is the result of such an interplay.

The metasystem paradigm is not a grand paradigm but may be considered a metaparadigm. It is worth considering in a system net-working context, as providing a rationale for determining the next system level. Depending on the directions of the contextual dimension considered, multilevels may extend in many possible directions, thus resulting in a heterarchy of systems, not a hierarchy. Each heterarchy provides a metalanguage for conflict resolution among the constituent systems for the specific system context. The rational bind of the systemic deductive or inference processes prevents a uniform guarantee of truth at that level. Through the metasystemic process, metarational considerations arise, leading to acceptance of otherwise contradictory assertions. We might say that wisdom is gained through a dialectic process of resolving paradoxes and metaparadoxes.

Such a dialectic process is an important component for the establishment of a metasystemic structure, even though it may result in approximate communication among the constituent systems. In this sense it differs from the multilevel systems theory of Mesarovic[5] which accepts the levels as given. The very possibility of repeated voids in the dialectic, will induce the constituent systems to seek the establishment of a metasystemic level.

Further research could clearly develop and extend these concepts. Policy sciences are full of metasystemic considerations. A formal approach of defining metasystems from a given family of systems, could yield metarational outcomes for resolving intersystemic conflicts and for system formation. At another level, the metasystem paradigm could give rise to such concepts as

> *meta-inputs, meta-outputs, meta-feedback, meta-feedforward, meta-feedwithin, meta-conflict, meta-consensus, meta-controllability, meta-observability, meta-centralisation, meta-decentralisation,* and so on.

At another level, we could seek the connection between the four types of *rationality* proposed in this book (van Gigch, Chapter 18) and the concept of *metarationality* (van Gigch, Chapter 19). The latter concept is certainly *structural* in that an outcome is considered metarational if a corresponding metaoutcome in the metagame is rational. It is *evaluative*, as it deals with the outcome preferences (ends); it is *substantive*, as it asks if sanctions are credible in the case of symmetric metarational outcomes or if the conflicting outcome is made stable through preference changes. It is also *procedural*, in that many different types of metagames are possible and thus result in not identical metarational outcomes.

An organisation is held together due to co-operative decision processes, for example, lower level decisions (managerial or otherwise) are adapted to higher level decisions (executive level or a total problem solving level). Unresolved conflicts in organisations are symptoms of decision-making failures. An application of the metasystem paradigm concentrates on evaluating the extent of these unresolved conflicts as a diagnostic for the severity of system formation problems.

By recursion, the paradigm can be extended to network systems in intersystemic contexts. Essential information and intelligence at the top echelons consist of metaintersystemic considerations. Furthermore, system failures must be considered in the dynamic ebb and flow of organisational decision making. There exists a certain dynamic to the forces of centralisation (which leads to the formation of a metasystem level) and the forces of decentralisation (which leads to the removal of the metasystem level). What may appear as a system failure according to one perspective, may in fact be a phase in the continuing dynamic process of system formation and reformation (or restructuring) (Capra's[1] Chapter 15: *The Cosmic Dance*).

Designing intelligent organisational information systems with advances in computer and information technology is another area where the metasystem paradigm could be used to provide system dysfunctional diagnostic indicators. In distributed information systems, 'peer systems' are designed to resolve conflicts.[6] There are no priority hardware or software configured 'meta systems'. The system as a whole evolves such a procedure.[7] Multimodelling concepts are essential and, in fact, a metadesign paradigm can be useful in that the specific activities of local user groups are not known. Networking Information Systems in multiechelon conglomerates or multichain systems provide areas where the metasystem paradigm could be tested, refined and evolved.

REFERENCES

1. F. Capra, *The Tao of Physics* (Bantam Books, New York, N.Y., 1976).
2. B. H. Voorhees, *Systems, Models, Methods: Epistemic Instrumentation*, Chapter 7, infra.
3. D. R. Hofstadter, *Gödel, Escher, Bach: An Eternal Golden Braid* (Vintage Books, New York, 1979).
4. R. K. Ragade, A Mathematical Model of Approximate Communication in Information Systems, *Proceedings of the 22nd. Annual Meeting of the Society for General Systems Research*, (Denver, Colorado, 1977). pp. 334–346.
5. M. D. Mesarovic, D. Macko and Y. Takahara, *The Theory of Hierarchical Multilevel Systems*. (Academic Press, New York, 1970).
6. M. Davis, *Distributed Systems: Architecture and Implementation*. (Springer Verlag, New York, 1981).
7. L. A. Belady and M. M. Lohman, The Characteristics of Large Systems, In: *Research Directions in Software Technology*, (MIT Press, Cambridge, Mass., 1979).

INDEX

Author references are in *Italics*